The Future of Land-Based Strategic Missiles

The Future of Land-Based Strategic Missiles

Edited by: **Barbara G. Levi, Mark Sakitt, Art Hobson**

This volume was prepared by a study group of the Forum on Physics and Society of The American Physical Society. The American Physical Society has neither reviewed nor approved this study.

Library of Congress Cataloging in Publication Data

The Future of land-based strategic missiles.

1. Intercontinental ballistic missiles. 2. United States--Military policy. 3. Strategic forces--United States. 4. Nuclear weapons--United States. I. Levi, Barbara G. II. Sakitt, Mark. III. Hobson, Art, 1934–

UG1312.I2F88 1989 358'.1754 89-6510

ISBN 978-0-88318-619-0

Contents

Preface

The United States has long debated the future of land-based missiles but has still not named a successor to the current Minuteman intercontinental ballistic missile force. About 50 MX missiles have found at least temporary homes in old Minuteman silos, while developments on other options proceed at various paces. It is time for a decision.

This book is intended to promote informed debate rather than to advocate any particular missile system. It presents background material on many options that are still considered viable, and evaluates each according to a consistent set of criteria.

The authors of this study endeavored to present a more coherent report than is sometimes achieved in collections of individually authored papers. Thus Part I begins with a list of "findings" on which all participants agree.

Parts II and III put land-based missiles into their historic and strategic context. Part III defines the viewpoint for evaluation of *all* land-based missile options: First one must decide the *purpose* that our nuclear forces are intended to serve; the system selected will depend strongly on the answer to that question, and the question itself is still widely debated. Then one must examine *which properties* of a missile system enable it to best fulfill the intended purpose. All land-based missile systems then must be evaluated in this light.

Part IV contains summary evaluations of each option. The authors endeavored to evaluate each option with respect to each of the following: survivability, stability, lethality, command and control, verifiability, and cost. They furthermore assessed these properties as a function of the future nuclear arsenals that might exist under four possible arms control futures: (1) no arms control constraints on nuclear arsenals, (2) constraints consistent with the SALT II accord, (3) limitations proposed for START, and (4) "finite deterrence," that is, deep reductions to about 2000 strategic warheads. Each of these evaluations was produced primarily by the person who had written a backup chapter on the corresponding missile option, but each was critiqued by other members of the study group, and a consensus was reached on all evaluations.

Part V consists of twelve background papers prepared by individual participants. All were reviewed by other members of the study group, but none claims to represent any viewpoint other than that of its author(s).

The options examined in this study are as follows:

•*No upgrade of land-based missiles.* Continued reliance on the silo-based Minuteman and MX missiles.

•*Launch on warning and launch under attack.* Adoption of a policy to launch missiles when either warning or confirmation of an attack is received.

•*Terminal defense of silos.* Defense of silo-based missiles against incoming ballistic missiles by means of ground-based interceptors.

•*Strategic diad.* Reliance only on the sea- and air-legs of the strategic triad.

•*Land-mobile Midgetman.* Deployment of single-warhead missiles on hardened mobile carriers.

•*Rail-garrison MX.* Deployment of MX missiles on railroad trains housed in military garrisons.

•*Midgetman or MX in superhard silos.* Construction of missile housing that is about 50 times as resistant to destruction as current silos.

•*Bunkered mobile.Midgetman.* Deployment of Midgetman missiles on carriers that can dash to a multiplicity of shelters, either vertical or lateral "semi-silos."

•*Multiple silos.* Deceptive deployment of a few hundred Minuteman, MX, or Midgetman warheads in a few thousand silos of considerable hardness.

•*Deep underground basing.* Burial of missiles deep underground, to be dug out after an attack.

The study group began under the auspices of the American Physical Society's Forum on Physics and Society. Peter Zimmerman, now of the Carnegie Endowment for Peace, and Herb Nelson of the Naval Research Lab, initiated the study and eventually attracted to it the ten physicists who have collectively authored this book. The Forum encouraged the effort and provided travel funds for several joint meetings of the participants but the participants volunteered their own time to research and write on the topics reported here. This book is the responsibility of the study group alone: neither the Forum nor the American Physical Society reviewed the final product.

Studies like this would be impossible without back-up assistance. The editors thank Marie Riley of Megabyte, Inc., of Fayetteville, Arkansas, for handling innumerable word processing questions and for other computer assistance, and Jean Eaton and Barbara Caudle of the University of Arkansas Physics Department for invaluable secretarial help.

The study group sincerely hopes that its efforts to collect and assess publicly available information on these options will promote rational and informed debate concerning them.

Paul Craig
David Hafemeister
Art Hobson
Ruth H. Howes
Barbara G. Levi
John Michener
Mark Sakitt
Leo Sartori
Valerie Thomas
Peter D. Zimmerman

About the study group

Paul P. Craig is a professor in the Department of Applied Science, University of California, Davis, and faculty associate at the Lawrence Berkeley Laboratory. He has a B.S. in physics from Haverford College and a Ph.D. from Cal Tech in cryogenics. He did basic research (unclassified) at Los Alamos during 1959-62, was at Brookhaven National Laboratory during 1971-75 and has been at the University of California since 1975. He was a member of the American Physical Society's Panel on Public Affairs in 1972-73 and has served on the Board of Directors and the Executive Committee of the Environmental Defense Fund. He was vice-chair, and then chair, of the Forum on Physics and Society during 1985-87. He is coauthor of the text *Nuclear Arms Race: Technology and Society.*

David Hafemeister is a professor of physics at California Polytechnic State University. He was a science advisor in the US Senate during 1975-77, special assistant to an Under Secretary of State during 1977-78, visiting scientist in the Office of Nuclear Proliferation Policy of the Department of State during 1978-79. He was at MIT during 1983-84, the Lawrence Berkeley Laboratory during 1985-86, the US State Department's Office of Strategic Nuclear Policy in 1987, Stanford in 1988, and Princeton in 1989. He has co-edited *Arms Control Verification*, and *Physics, Technology and the Nuclear Arms Race*, and *Nuclear Arms Technologies in the 1990s*. He was chair of the Forum during 1985-86.

Art Hobson is a professor of physics at the University of Arkansas, Fayetteville, where he has worked since 1964. He obtained his physics Ph.D. in theoretical statistical mechanics from Kansas State University. He has published numerous papers in his physics research field and authored a research monograph, *Concepts in Statistical Mechanics* (Gordon and Breach, New York, 1971). He is the author of *Physics and Human Affairs* (Wiley, New York, 1982), a physics textbook for non-scientists that also presents historical, philosophical, and social perspectives, including chapters on nuclear war and nuclear power. For many years he has taught a large-lecture course entitled "Physics and Human Affairs," and he developed and teaches an interdisciplinary course entitled "Peace and the Nuclear Arms Race." He spent most of 1985 as a visiting researcher at the Stockholm International Peace Research Institute. For the past two years he has served as editor of the Forum's newsletter *Physics and Society*.

Ruth H. Howes is a professor of physics and astronomy at Ball State University where she serves as director of the Center for Global Security Studies. She holds a B.A. in physics from Mount Holyoke College and an M.S. and Ph.D. in physics from Columbia University. Before coming to Ball State University, she taught at the University of Oklahoma and Oklahoma City University. She has taught courses on the nuclear arms race since 1976. In 1984-85 she served as a William C. Foster Fellow in the Bureau of Verification and Intelligence of the Arms Control and Disarmament Agency. Her research interests also include the application of nuclear physics to problems in geology, archaeology, and art history. In 1986, she was named speaker of the year by the Indiana Academy of Science.

Barbara G. Levi is currently a visiting professor of physics at Rutgers University. She has a B.A. in physics from Carleton College and a Ph.D. in high-energy physics from Stanford University. After receiving her degrees, she taught physics at Fairleigh Dickenson University from 1970 to 1976 and at Georgia Tech from 1976 to 1980. From 1980 to 1987 (except for one year spent at Bell Labs) she was a member of the research staff at the Center for Energy and Environmental Studies at Princeton University. Her research at Princeton focused principally on energy and nuclear weapons policy. She coauthored two articles in *Scientific American* based on this work. Dr. Levi has written research news articles for *Physics Today* for nearly 20 years, and has consulted for the Congressional Office of Technology Assessment on such studies as *Nuclear Proliferation and Safeguards*. She was the 1988/89 chairman of the APS Forum on Physics and Society.

John Michener is vice president for engineering of Indata Inc. Until recently he was a senior research scientist for the Siemens Corporate Research Laboratories. He obtained his Ph.D. in 1984, from the Materials Science Program of the Mechanical Engineering Department at the University of Rochester. He holds an MS in Materials Science from the University of Rochester and a BS in Physics from the University of Maryland. He has published in cryptography, defense analysis, electronic devices, materials science, and cirucit testing. He holds patents in electronic devices, electrophotography, and cryptoography. In defense analysis he has studied countermeasures to orbital defense systems, and the protection of land based missiles. In cryptography, he has studied techniques for microprocessor implementations and the novel application of modern electronic devices for cryptographic protection of data and communications.

Mark Sakitt is a senior scientist in the Physics Department at the Brookhaven National Laboratory. He has a Ph.D. in high energy physics from the University of Maryland. He is a visiting professor at the State University of New York at Stony Brook where he has been teaching a course on "Technology and Policy in National Security," and is currently also teaching a course on "Proliferation of Nuclear Weapons." During 1986-87 he was a Carnegie Science Fellow at the Center for International Security and Arms Control at Stanford University. He is the author of *Submarine Warfare in the Arctic: Option or Illusion* (Center for International Security and Arms Control, Stanford CA, 1988) critiquing current US naval strategy. He is a Fellow of the American Physical Society and has served on the Forum's Executive Committee.

Leo Sartori is a professor of physics and of political science at the University of Nebraska at Lincoln. He has taught at Princeton, from 1972 until 1978. During 1978-81 he served at the Arms Control and Disarmament Agency, and in 1979 was senior ACDA advisor to the US SALT delegation in Geneva. He has been a visiting scientist at the Arms Control Centers at MIT and Stanford, and is spending the 1988-89 academic year at the Office of Arms Control, Department of Energy. He was vice chair and chair of the Forum on Physics and Society, 1983-85. He chaired the committee which initiated the Forum-sponsored studies, of which the present volume is the second.

Valerie Thomas is a research associate at the Center for Energy and Environmental Studies at Princeton University. She has a B.A. in physics from Swarthmore College and a Ph.D. in theoretical high energy physics from Cornell University. She was a research fellow in the Program on International Peace and Security at Carnegie Mellon University, 1987-88, and is currently a member of the joint research project on disarmament of the Federation of American Scientists and the Committee of Soviet Scientists Against the Nuclear Threat.

Peter D. Zimmerman is a senior associate at the Carnegie Endowment for International Peace, Washington, D.C. He holds a B.S. in physics from Stanford, an M.S. from the University of Lund in Sweden, and his Ph.D. in nuclear and particle physics from Stanford. He was a member of the physics faculty at Louisiana State Univertsity during 1974-87, during which time he was a visiting researcher at DESY in Germany, at UCLA, at Fermi Lab, at the University of California at San Diego, and at Princeton University. He was a William C. Foster Fellow with the US Arms Control and Disarmament Agency during 1984-86, during which time he spent a year as a technical advisor to the US START delegation in Geneva. He is currently directing the project on SDI technology and policy at the Carnegie Endowment, and co-directing the project on commercial observation satellites and national security. He served as secretary-treasurer of the Forum on Physics and Society during 1984-88.

Acronyms and glossary

ABM	an anti ballistic missile, a surface-based missile designed to shoot down an incoming missile or warhead
ABM Treaty	A treaty that resulted from the SALT I negotiations, severely limiting defenses against ballistic missile attack
acceleration	any change in the speed or direction of motion of an object
airblast	expanding wave of high-pressure air created by the heat of an exploding weapon
airburst	an explosion of a nuclear weapon at some distance above the ground, *cf. groundburst*
ALCM	air-launched cruise missile, *see cruise missile*
ASAT	anti-satellite weapon
atm	a unit of air pressure, one atm is normal atmospheric pressure, about 15 pounds per square inch, or about 10^5 newtons per square meter
B-1	new low-flying US strategic bomber having low radar cross-section and some capability to overfly the Soviet Union
B-2	possible future US strategic bomber having very low radar cross-section, also known as the "Stealth" bomber
B-52	older US strategic bomber whose main strategic role would be as a "stand-off" launcher of air-launched cruise missiles from outside the Soviet Union
ballistic missile	a missile that is "thrown" (like a ball) at its target, powered during the first minutes of its flight by rocket engines, *cf. cruise missile*
barrage	an attack in which warheads are directed to cover a chosen area with destructive blast, as opposed to an attack targeted at one or more specific points
blast	*see airblast*

BMD	ballistic missile defense, nearly synonomous with (but also including space-based defenses) ABM defense
boost stage	the portion of a ballistic missile's journey during which it is being accelerated by rocket engines
breakout	the sudden and massive violation of an arms control treaty
bus	the maneuverable platform that carries ballistic missile warheads during the post-boost phase, and that maneuvers and releases each warhead toward its target
C3	command, control, and communications
CEP	the radius of the circle of equal probability, i.e. the distance from a target within which there is a 50% chance of a warhead directed at that target exploding, a quantitative measure of inaccuracy
closely-spaced basing	an arrangement of missile silos in which the silos are so close together that attacking warheads would destroy each other in the process of attempting to destroy the silos
compressive strength	the stress (force per unit area) that a solid body under compression can withstand without fracturing or otherwise failing
control	usually synonomous with command, control, and communications; used in this book in reference to nuclear forces
cooperative measures	ways of verifying arms agreements that depend on the (agreed-upon) cooperation of the other side, e.g. on-site inspection
crater radius	the radius of the inner bowl of a nuclear crater, *cf. inner bowl*
crisis stability	a nuclear balance that offers no incentives for either side to strike first in a crisis
cruise missile	a missile that flies through the air to its target, powered by a jet engine, *cf. ballistic missile*
CSB	*see closely spaced basing*
dash-mobile basing	Midgetman deployment plan in which the mobile launchers would be parked in vulnerable bunkers next to Minuteman silos, ready to dash on warning of nuclear attack

deep underground basing	plan to base ICBMs deep underground where they will be protected from nuclear blasts
deploy	to put into the field, ready for military action
depressed trajectory	a ballistic missile flight path that is lower and faster than the "normal" minimum-energy path, used especially in reference to SLBMs
designated deployment area	a region to which a particular type of weapon is restricted by arms control agreement
destabilizing	causing reduced stability, *cf. stability*
deterrence	the avoidance of nuclear war by the threat of retaliation from any side that is attacked
DEW	*see directed energy weapons*
DEW line	distant early warning radar line in Canada
diad	strategic nuclear forces containing only two members of the triad existing today, *cf. triad*
directed energy weapon (DEW)	any laser or particle beam weapon that acts at, or nearly at, the speed of light
DOD	Department of Defense
dynamic blast	*see dynamic pressure*
dynamic pressure	a briefly-applied pressure (force per unit area), *cf. static pressure*
earth penetrator	a warhead that penetrates some distance into the earth and then explodes
electromagnetic pulse	sudden burst of electromagnetic energy sent out by a nuclear explosion, harmless to biology but destructive to electrical equipment especially solid-state electronic devices
electromagnetic wave	a wave of electromagnetic field energy, including radio, infrared, visible light, ultraviolet, x-ray, gamma ray
EMP	*see electromagnetic pulse*

equivalent megaton	combination of warheads and yields that can barrage the same area (with the same effectiveness) as can be barraged by a single one megaton warhead; a single warhead of yield y (megatons) has equivalent megatonnage $y^{2/3}$
ERIS	exoatmospheric reentry vehicle interception system, an ABM device that attacks incoming missiles above the atmosphere
finite deterrence	*see minimum deterrence*
fratricide	the destruction of a nuclear weapon as it approaches a target, due to effects from other weapons exploding in the vicinity
gamma ray	a very high frequency electromagnetic wave, emitted by a nuclear explosion and other nuclear reactions
giga (G)	one billion, 10^9
GLCM	ground-launched cruise missiles, intermediate-range cruise missiles deployed in Europe and now banned by the INF Treaty
GPS	*see NAVSTAR*
groundburst	an explosion of a nuclear weapon at ground level, *cf. airburst*
guidance	the process by which a missile controls its flight in order to hit the target
GWEN	ground-wave emergency network, a land-based communication system hardened against EMP and other nuclear effects
hard target	a target that is strengthened (up to some limit) against nuclear blast, also used in reference to protection against other nuclear effects such as radiation
hardened mobile launcher (HML)	the mobile carrier for Midgetman
HEDI	high endoatmospheric defense interceptor, an ABM device that attacks incoming missiles inside the atmosphere
high frequency (HF)	electromagnetic radio waves in the 3-30 megahertz (MHz) range
HML	*see hardened mobile launcher*
ICBM	intercontinental ballistic missile, *see ballistic missile*

independently survivable	a strategic force (either the ICBMs, or SLBMs, or bombers) that is invulnerable, i.e. that cannot be effectively destroyed by any plausible attack, *cf. synergistically survivable*
inertial guidance	a missile (or other, e.g. airplane) guidance system that measures accelerations (i.e. deviations from inertial, or unaccelerated, motion) and from that calculates the velocity and position of the missile
INF Treaty	Intermediate Nuclear Forces Treaty signed in 1987, eliminates two range-categories of medium-range ballistic and cruise missiles
inner bowl	the central portion of a nuclear crater, in which the ground has been dug out by the explosion, *cf. outer bowl*
intelligence cycle time (ICT)	the average time required for a surveillance satellite to first become aware of some ongoing event on the ground on the other side and then to communicate that fact to the satellite's command
kilo (K,k)	one thousand, 10^3
kiloton (kT)	a measure of explosive power, the energy released in the explosion of one thousand tons of TNT, about 10^{12} calories or 4.2×10^{12} joules.
Krasnoyarsk	site of a Soviet radar that (according to all Western observers) violates the SALT I Treaty
kT	*see kiloton*
L	*see lethality*
launch on warning (LOW)	the launch of nuclear weapons on the warning that the other side's nuclear weapons have been launched
launch under attack (LUA)	the launch of nuclear weapons immediately upon the explosion of the other side's first nuclear weapons on one's home country, often used synonomously with launch on warning
launcher	a vehicle that launches one or more warheads toward their targets, i.e. an ICBM or SLBM or bomber
LCC	launch control center, installation where military personnel send out the electromagnetic signals to launch nuclear weapons

lethality (L)	for a warhead, the combination $y^{2/3}/CEP^2$, a measure of the warhead's ability to destroy hardened targets
LOW	*see launch on warning*
low frequency	electromagnetic radio waves in the 3-300 kilohertz (KHz) range
LPAR	large phased array radar
LUA	*see launch under attack*
maneuverable reentry vehicle	*see MaRV*
MaRV	maneuverable reentry vehicle, an RV that maneuvers aerodynamically in the atmosphere as it approaches its target for the purpose of either evading the other side's ABM devices or of accurately hitting the target
mega (M)	one million, 10^6
megaton (MT)	a measure of explosive power, the energy released in the explosion of one million tons of TNT, about 10^{15} calories, or 4.2×10^{15} joules.
Midgetman	proposed US small ICBM, single-warhead, lightweight
Milstar	a planned system of hardened communication satellites
minimum deterrence	a condition in which both superpowers have the minimum number of nuclear weapons needed to ensure deterrence (usually 1000-2000 warheads)
Minuteman (MM)	*see Minuteman II and III*
Minuteman II (MM II)	US single-warhead silo-based ICBM
Minuteman III (MM III)	US 3-warhead silo-based ICBM
Minuteman IIIA (MM IIIA)	advanced version of Minuteman III, has a higher-yield warhead, lethal against Soviet silos
MIRV	multiple independently-targetable reentry vehicle, a ballistic missile carrying more than one warhead and which can release each warhead toward a different target

MM	*see Minuteman II and III*
MT	*see megaton*
MX missile	US 10-warhead missile, currently in silos
national command authority (NCA)	the President or his designated successor having authority over nuclear weapons
national technical means (NTM)	ways of verifying arms agreements that do not depend on the cooperation of the other side, e.g. satellite surveillance
NAVSTAR	the system of "global postitioning satellites" (GPS) satellites that can determine a receiver's location extremely accurately
NCA	*see national command authority*
NEACP	National Airborne Command Post, an aircraft maintained on alert just outside Washington, from which the NCA could command a war
negative control	nuclear weapons controls that prevent their unauthorized use, *cf. positive control*
neutron pulse	the burst of neutrons sent out from a nuclear explosion
NTM	*see national technical means*
on-site inspection	one type of cooperative arms control measure, in which inspectors from one side are present at the installations of the other side
OTA	the US Office of Technology Assessment
outer bowl	the outer portion of a nuclear crater, where the air pressure from the explosion has pressed the ground down below its normal level,*cf. inner bowl*
overpressure	the excess air pressure, above normal atmospheric pressure, produced by an exploding warhead
penetration aids	devices to enable attacking warheads to overcome the defenses of the other side
permissive action link (PAL)	a coded switch on a nuclear weapon that makes launch of the weapon dependent upon possession of the proper code from higher authorities

Pershing II	an intermediate-range ballistic missile deployed in Europe, now banned by the INF Treaty
pindown	a series of high-altitude nuclear explosions over an ICBM base, designed to prevent the ICBMs from launching
positive control	nuclear weapons controls that ensure their use when authorized, *cf. negative control*
post-boost stage	the portion of a ballistic missile's journey immediately following the boost stage, during which the missile's "bus" is maneuvering for greater accuracy and releasing its individual warheads toward the targets, *cf. bus*
pre-empt	to initiate nuclear war by attacking first in a crisis, especially when it is feared that the other side is about to attack
price to attack	the ratio of the number of warheads used by the attacker to the number destroyed in a nuclear attack, inverse of the warhead-exchange ratio
propellant	the solid or liquid fuel that propels a rocket
PS	probability that a given target will survive when attacked by a single warhead, taking warhead reliability into account, *cr. SSPS*
psi	pounds per square inch, a unit of pressure, equals about 1/15 atm (1 atm is about 15 psi).
r, R	reliability, sometimes radius, *see reliability*
rad	radiation absorbed dose, 1 rad is that amount of radiation which deposits 0.01 joule of energy into one kilogram of absorbing material, about 500 rads will kill a person
radar	radio detection and ranging, equipment that determines the location of objects by emitting radio waves and then detecting the return waves after their reflection from the distant object
radius of destruction (RD)	maximum distance from an exploding warhead at which a given target can be destroyed
rail-based MX	see rail-garrison MX
rail-garrison basing	a plan to base MX missiles on trains that will be parked ("garrisoned") in bunkers on military bases, ready to dash when alerted

random-mobile basing	Midgetman deployment plan in which the mobile launchers would move continuously and unpredictably over large tracts of military land in the southwest, and could dash onto larger tracts upon warning
RD	*see radius of destruction*
reconnaissance	military surveillance, for intelligence purposes
reentry vehicle (RV)	the heat shield (and usually its warhead and other contents) surrounding a ballistic missile explosive warhead, to protect the warhead during reentry into the atmosphere
reinforced concrete	concrete made more indestructible by means of imbedded metal wires or ribbons or bars
reliability (r, R)	the reliability of an ICBM warhead is the probability that it will be properly launched and will explode as planned near its target, i.e. the probability that it is not a dud.
RV	*see reentry vehicle*
S	*see strength*
SAC	Strategic Air Command
SALT I	talks held during the early 1970s that led to an interim agreement limiting strategic weapons and to the ABM Treaty, *cf. ABM Treaty*
SALT II Treaty	treaty signed but not ratified during late 1970s, putting a numerical upper limit on the strategic nuclear weapons
Scowcroft Commission	issued an influential report, in 1983, on land-based missiles
SDI	*see Strategic Defense Initiative*
silo	hardened underground facility for protection and launch of an ICBM
single-wave attack	an attack in which only one warhead is directed at each target, *cf. two-wave attack*
SLBM	submarine-launched ballistic missile
SLCM	sea-launched cruise missile, *see cruise missile*

small ICBM	any lightweight single-warhead ICBM, often used synonomously with Midgetman ICBM
SPOT	a French commercial photographic satellite
SRAM	short-range attack missile, launched from bombers as a penetration aid for suppression of enemy air defenses, range usually under 150 kilometers.
SS-18	Soviet 10-warhead silo-based ICBM
SS-19	Soviet 6-warhead silo-based ICBM
SS-20	a Soviet intermediate-range ballistic missile, now banned by the INF Treaty
SS-24	new Soviet rail-based 10-warhead ICBM
SS-25	new Soviet land-mobile small ICBM, single-warhead, about the weight of the US Minuteman ICBM
SSBN	a submarine carrying SLBMs, nuclear powered
SSKP	single-shot kill probability, the probability that a single exploding warhead (not a dud) will destroy its target, 1-SSPS
SSPS	single-shot probability of survival, the probability that a particular target will not be destroyed by a single exploding warhead (not a dud), 1-SSKP
stability	*see crisis stability*
START	Strategic Arms Reductions Talks, ongoing negotiations that are focused on a proposed agreement to reduce superpower strategic ballistic missile warheads by about 50%, and superpower strategic weapons by 30-40%
static pressure	an unchanging pressure (force per unit area), *cf. dynamic pressure*
Stealth bomber	*see B-2*
Strategic Defense Initiative (SDI)	A program to defend the US against nuclear attack by destroying the missiles and/or warheads before they explode at their targets
strategic triad	*see triad*

strategic weapons	weapons designed for long-distance attack on the infrastructure (industry, population, military infrastructure, strategic weapons) of a country
strength (S)	in reference to a target, the maximum overpressure that the target can sustain without being destroyed, also called "hardness," usually measured in atm
stress	the pressure, or force per unit area, acting in a particular direction within or at the surface of a solid body
superhardened silo	in this book, a silo that is so strong that it can only be destroyed by being included in the crater of a nuclear explosion; also used more generally elsewhere to mean any target that is much stronger than current Minuteman silos
surprise attack	to initiate strategic nuclear war by attacking first when there is no crisis, *cf. pre-empt*
synergistically survivable	two forces (especially bombers and ICBMs) that cannot both be destroyed in a single attack, even though either one might be individually destructable, *cf. independently survivable*
TEL	transporter-erector-launcher for carrying and launching land-mobile ballistic missiles, especially the US Pershing II and the Soviet SS-25
tensile strength	the stress (force per unit area) that a solid body under tension can withstand without fracturing or otherwise failing
terminal guidance	any guidance applied to a missile during the final phase of its flight, as it approaches its target
throw-weight	the "useful" weight which is placed on a path toward the target by a ballistic missile's booster rockets, usually includes all the missile's RVs plus the "post-boost vehicle" that releases the individual RVs plus any penetration aids
tonne	a metric ton, 1000 kilograms, about 2200 pounds
triad	the threefold strategic nuclear forces (ICBMs, SLBMs, and bombers) of both superpowers
Trident I	US 8-warhead SLBM
Trident II	planned US 8-warhead SLBM, more accurate and with larger yield than Trident I

Trident submarine	new US SLBM-carrying submarine, can carry Trident I or II missiles
two-wave attack	an attack in which two warheads are directed in succession at each target, *cf. single-wave attack*
very high frequency (VHF)	electromagnetic radio waves in the 30-300 megahertz (MHz) range
warhead	an explosive device, often used synonomously with "nuclear weapon," often includes also the conical heat-shield surrounding the explosive device
warhead-exchange ratio	the ratio of the number of warheads destroyed to the number used by the attacker, in a nuclear exchange, inverse of the price to attack
y, Y	*see yield*
yield (y, Y)	explosive energy release, in this book usually measured in megatons

Part I.

Study findings

Paul Craig, David Hafemeister, Art Hobson, Ruth H. Howes, Barbara G. Levi, John Michener, Mark Sakitt, Leo Sartori, Valerie Thomas, Peter D. Zimmerman

General findings

1. No land-based missile option considered in this study simultaneously optimizes survivability, stability, cost, verifiability, command and control, and flexibility.

Deciding which new missile system to deploy is tantamount to deciding which missions the missile should perform and which properties have highest priority. Any decision forces a tradeoff among various, and often conflicting, priorities. See Part III of this study.

2. The land-based systems evaluated in this study differ considerably when evaluated in terms of various properties.

Superhard silos are not vulnerable to rapid escalation or even to a bolt-out-of-the-blue attack. This fixed basing mode simplifies verification and offers reasonably secure command and control. However, it is expensive and may become vulnerable if warheads of greater accuracy and/or higher yield are deployed.

Multiple-shelter basing for MX, Midgetman, or Minuteman III exacts a high price to attack without relying on warning time, and allows both lower operational costs and better command and control than do mobile modes. The cost of this mode varies with the missile chosen. The degree of protection against attack is not yet fully explored.

Land-mobile Midgetman is survivable if given 30 minutes tactical warning, and survival would not be significantly affected by warheads of greater accuracy.

This option carries a high cost, perhaps $40 billion, and would complicate the tasks of verification and command and control.

Rail-mobile MX is survivable if given three hours strategic warning. It is the cheapest and operationally the simplest of the mobile modes. Verification and command and control are more difficult than with silo basing. Both because it has ten warheads and because it has high vulnerability to short-warning attacks, the rail-mobile MX could decrease crisis stability.

3. None of the basing modes can remain invulnerable unless constraints are placed on technical improvements or numerical increases in weapons and delivery vehicles.

Today's Minuteman and MX silos are vulnerable to destruction by accurate Soviet ICBMs. These missiles might be placed in superhard silos, some 50 times more blast-resistant than current silos, but even these structures could be threatened by high-yield Soviet warheads with accuracies attainable by the mid 1990s. Superhard silos would also be vulnerable to terminally guided, maneuverable reentry vehicles (MaRVs) or earth-penetrating warheads.

Land-mobile or rail-mobile missiles, which rely for survival on dispersal, could be destroyed by a large barrage attack if the Soviets deployed enough missiles. Mobile missiles are also vulnerable to short-warning attacks from missiles launched on depressed trajectories from off-shore submarines. More speculatively, the Soviets might conceivably shorten the intelligence cycle time to the point where satellites could track mobile missiles from space and relay information on their position in real time to MaRVs.

Point-site defense of US missiles might be defeated by Soviet penetration aids or direct attacks on sensors, or simply overwhelmed by increased numbers of attacking warheads.

In short, continued survivability of future land-based missile systems cannot be assured simply by design of new basing modes but requires attention to trends in offensive capabilities. Continued survivability may require such arms control measures as ceilings on the numbers of new missiles, bans on testing and deployment of maneuvering or penetrating warheads, and restrictions against testing missiles in depressed trajectories.

In Part V see Chapter 3 on current vulnerability, Chapter 7 on land-mobile Midgetman, Chapter 8 on rail-mobile MX, and Chapter 11 on superhard silos, .

4. One option is not to upgrade the land-based missile force at all.

No new land-based missiles should be deployed, especially in this time of tight budgets, unless they clearly offer greater security both now and in the long term. Proposed upgrades to the land-based missiles have been motivated by perceived needs for survivability against nuclear attack, stability in a crisis, and more modern capabilities. Before the US selects any new land-based system it must clearly evaluate these needs and ask whether they are met by the proposed new system. At the same time, the Minuteman missiles are nearing the end of their service life and some replacement may be mandated if the triad is to be maintained.

5. If the land-based missiles were not upgraded, the US strategic arsenal might evolve toward a diad as existing missiles age.

Some of the capabilities formerly provided uniquely by land-based missiles are now being acquired by the other legs as well. Trident II submarine-launched ballistic missiles, as well as cruise missiles launched from surface ships, attack submarines, and strategic bombers, are expected to have accuracies approaching those of the most modern ICBMs.

Land-based missile survivability seems an elusive goal, yet the other two legs of the triad continue to have high survivability.

On the other hand, land-based missiles may be needed as a hedge against future vulnerability of the other two legs of the triad. They also have become a symbol of nuclear prowess, possibly important in international relations and arms control negotiations.

If flexible response were not required and one only needed simple deterrence (not now the current US doctrine) then a diad would be satisfactory.

See Part V Chapter 5.

6. New land-based missiles could increase US ICBM "counterforce" capability to attack missile silos and other similarly hardened targets.

To threaten ICBM silos and hardened command posts requires the accuracy and power typically provided by ICBMs. Midgetman and MX are more accurate than the present Minutemen. The planned Trident II submarine ballistic missile will be able to destroy silos, but underwater sea-based forces cannot be as promptly and reliably commanded as can land-based forces.

7. Even if deployed in vulnerable basing, single warhead missiles contribute significantly to crisis stability, especially if there are arms control limits on the numbers of deployed warheads on each side.

It takes two warheads to confidently attack a silo-based missile, and more to destroy a missile in random motion or shuttling among multiple protective shelters. If each US missile carried just one warhead, the attacker would always have to spend more (several more, for mobiles) warheads than it destroyed. Thus, if a high fraction of both sides' land-based missiles had only one warhead each, any future confrontation would be more stable: even if nuclear war appeared inevitable, both sides would be less tempted to initiate conflict, for single warhead missiles lower the potential gain. See Part V Chapter 1.

8. Future US and Soviet arms control positions may be affected by land-based missile decisions

The START Treaty now being negotiated would allow both sides to keep most of their most modern forces, in roughly their present proportions. However, deeper reductions would require reapportionment of strategic forces. Especially with force levels of a few thousand warheads or less, flexibility and stability are enhanced by single-warhead missiles.

9. Missile basing modes should be considered no more survivable than their command, control and communications systems.

If missiles survive an attack but their command, control and communications (C^3) system does not, they are effectively disabled. C^3 survival is much easier if missiles are only required to execute retaliation on predetermined targets than if the missiles are meant to engage in an extended nuclear war. Even the appearance that C^3 systems are vulnerable could weaken deterrence by creating doubts that the US could respond to a first strike. See Part V Chapter 6.

Findings about specific options

1. Land-mobile Midgetman's high survivability and its stabilizing single warhead must be weighed against its several disadvantages.

Given constraints on numbers of land-based missiles, the land-mobile mode can be invulnerable. The price of this advantage is high, both in dollars and in operational complexity. Assuring adequate C^3 is more difficult than with silo-based missiles. High accuracy, for warheads launched from dispersed positions, probably requires many presurveyed launch sites. Security would also be more difficult, especially in a crisis, when the missiles must move from government to private land. Furthermore, the missile would not carry enough supplies and fuel to remain dispersed for long. If a START Treaty limits without prohibiting mobile missiles, the verification procedures may require cooperative measures beyond those required by the INF Treaty.

2. Off-shore submarines can attack mobile missiles with less warning time than can ICBMs. Thus mobile missiles might not solve the possible future problem of the simultaneous vulnerability of ICBMs and bombers to a large Soviet SLBM attack.

Mobile modes begin to dash after warning of an attack. While ICBMs reach US targets in 30 minutes, SLBMs arrive in about 15 minutes or even less if depressed trajectories are used. Because mobile missiles are much "softer" than silos, SLBMs are accurate enough to destroy them. Strategic bombers are similarly dependent on warning and share with mobile missiles a vulnerability to SLBM attack. The Soviets could thus conceivably destroy these two legs of the triad with their SLBM force. That scenario, however, requires the Soviets to deploy, either routinely or clandestinely, many submarines off our shores. The problem is more severe for dash-mobile Midgetman, in which missiles dash from known static locations, than for random-mobile Midgetman, in which the missiles are in nearly continual motion. See Part V Chapter 7 on land-mobile Midgetman and Part V Chapter 8 on rail-mobile MX

3. Mobility enhances survivability because missiles are hard to find, but complicates verification for the same reason.
An agreement to limit mobile missile deployments would require procedures for counting missiles in the field without knowing their individual locations. Such verification would require national technical means to be supplemented by cooperative measures. The INF Treaty has set important new precedents for such measures in the case that entire missile classes are banned. More elaborate procedures may be required for deployment of mobile missiles in restricted numbers. See Part V Chapter 2.

4. The rail-mobile MX could be deployed easily and inexpensively compared to other options, but is invulnerable only if it has about three hours to disperse from garrisons where it is based day-to-day.
The MX is the weapon with the greatest ability to threaten Soviet land-based missiles. With 20 warheads on each train and trains normally grouped in garrisons at known locations, this basing mode places many eggs in one basket. Dispersing trains requires them to move onto public land. Such early dispersal in a crisis has both stabilizing and destabilizing aspects. The rail-mobile MX is operationally simple and can remain dispersed for a reasonably long time. As with land-mobile missiles, assuring accuracy might require numerous presurveyed launch points. Communications lines along the tracks may be vulnerable to cratering detonations. Verification would require cooperative measures. See Part V Chapter 8

5. Multiple silo basing of MX, Minuteman, or Midgetman, might combine the survivability of the mobile modes with the operational ease of the fixed modes.
500 warheads on 50 MXs or 167 Minutemen or 500 Midgetmen might be moved deceptively between some 2500 shelters. Studies are needed to determine if it is possible to drill sufficiently strong silos into rock. Such a system would require much less land area than the similar "multiple aim point" MX basing mode discussed in the late 1970s. Attacking these multiple silos would require an implausibly large Soviet force, provided that the Soviets cannot find which silos contain the real missiles.

Detailed cost estimates are not yet available, but this mode could be cheaper than the land-mobile options if existing Minuteman missiles were used. Building the required silos for the larger MX missile would raise the cost.

6. Deep underground basing is very survivable but the missiles might not be able to emerge after an attack.
Deep underground basing would allow the US to respond to a nuclear attack only by striking predetermined targets. However, their ability to dig out after a nuclear attack might not be assured without atmospheric nuclear tests, which are banned. It is not clear whether deep underground basing is either feasible or invulnerable against a well-planned attack.

7. The vulnerability of missiles in any basing mode encourages a launch-on-warning or a launch-under-attack posture.
Either posture could be regarded as inexpensive operational answers to the vulnerability problem but they open the possibility of unauthorized, inadvertent,

or mistaken launching of nuclear missiles. This posture requires the US to rapidly assess an attack and to formulate and carry out an appropriate response, a very demanding requirement. Furthermore, hastily-launched US missiles might have limited usefulness if their targets had to be predetermined. The US could implement a scheme to arm or disarm a missile in flight, but must then worry about the security of the codes to accomplish this.

8. Deployment of terminal defense of nuclear missiles might provide partial survivability of ICBMs, but raises several arms control issues.

Terminal defenses look more promising today than they did when the US abandoned its option to defend missiles at its Grand Forks site. Radar and associated computer systems are more capable, new technologies may introduce novel components, and the defense can choose to defend only a limited number of silos. However, the expensive radars still remain vulnerable, and the offense can use many other countermeasures. Any decision to deploy terminal defense at a site other than Grand Forks would require renegotiation of the ABM Treaty. Furthermore some of the new technologies might create problems if they fall into the murky region between land-based and spaced-based defenses. See Part V Chapter 4 on hard point defense.

9. US missiles must not only survive enemy attack but must be able to penetrate defenses. If Soviet defenses improved significantly in the future, the US might counter this with penetration aids on its intercontinental missiles.

At this time we have seen no evidence that Soviet defenses are mature enough to affect the penetrability of US ICBMs. One response to possible future Soviet defenses is to develop penetration aids for existing weapons rather than proliferate new weapons. Such aids are currently under research and development in the US. See Part V Chapter 12.

Part II

Historical introduction

Paul Craig

Nuclear weapons have fundamentally altered the approach of the United States to national security. Advanced weapons and delivery systems have made the nation vulnerable - for the first time in our history - to direct, immediate and devastating attack. Hopes for defensive systems, if feasible, are on time scales of one or more decades, beyond the period of this study. For the foreseeable future we and the Soviets live at each others' sufferance. Necessarily, decisions on our weapons systems must take account of Soviet capabilities and intentions.

Nuclear weapons systems capable of penetrating to the heartland of the Soviet Union constitute the cornerstone of United States military strategic strategy. These weapons are placed on submarines, aircraft, and intercontinental ballistic missiles. Collectively these systems comprise the Strategic Triad. The structure of the Triad has evolved over four decades, since the beginnings of strategic nuclear planning in the late 1940s, when the Strategic Air Command (SAC) first acquired the capability of intercontinental delivery of nuclear weapons (1).

During the first decade of strategic nuclear weaponry the United States enjoyed unambiguous predominance over the Soviet Union. This predominance was not accompanied by any clear plan on how these United States nuclear weapons might be used. Although some discussion of the implications of nuclear weapons and delivery systems took place in the Truman Administration, it was not until the Eisenhower Administration that Secretary of State John Foster Dulles articulated the doctrine that the United States would respond to acts against the nation by massive retaliation "at a time and a place of our choosing".

Despite the forceful mode of expression and the apparent clarity of the Dulles doctrine, complications quickly emerged. The most important of these was the impossibility of articulating what sort of affront to the US would justify the use of weapons with the destructive power of the H-bombs that by then had been developed, or even of the more modest weapons of the size used at Hiroshima and Nagasaki. The Dulles Doctrine sounded good, but practical implementation proved elusive.

By the mid 1960s the Soviet Union had developed an extensive capability for delivering massive nuclear blows to the United States. The two nations became hostages to each other, in a situation which soon became known as Mutually Assured Destruction, or MAD. MAD was the result of the evolution of increasingly powerful military technology. It was not (and is not) so much a matter of policy as a fact of life.

Attempts to develop nuclear strategies must inevitably address the fundamental fact that we are affected by decisions made by other nations. This has led to analysis in terms of the so- called "rational decision maker." The idea is that deterrence will occur if, following a Soviet first strike, the United States can retain enough destructive capability to produce unacceptable damage to the Soviet Union. In classical theory of war (e.g. that of Clausewitz), conflict may occur if one side feels that by going to war there will be gains which exceed the losses. Alternatively, war can occur if a nation feels it is boxed in. Leaders may go to war in order to prevent an even worse disaster.

All these arguments assume that a potential enemy will undertake some sort of rational analysis of options. An enemy which does not operate this way is "irrational", and it is entirely unclear what forms of deterrence might work against such an enemy.

The MAD fact that our lives depend on Soviet decisions is deeply disturbing. Inevitably much effort is put into looking for ways out. The United States seems to prefer technical solutions, as reflected for example in our approach in the 1960s to MAD. We set out in the late 1960s to develop defense systems against Soviet missiles. The resulting antiballistic missile (ABM) program was hotly debated, and eventually recognized as technically infeasible then. This recognition by both the US and the Soviet Union led to the ABM treaty which entered into force on October 3, 1972. At the time that treaty was signed the United States was still far ahead of the Soviet Union in nuclear weapons and delivery systems. But the Soviet Union had enough deliverable nuclear capability to destroy us; this totally neutralized our advantage. The ABM treaty was a result of the recognition of two factors, both of which have been challenged by the Strategic Defense Initiative. First, effective defense against modern nuclear weaponry is futile for technical reasons and secondly, the United States and the Soviet Union both realized that defenses were likely to fuel a spiraling arms race.

The next era in the nuclear arms race, lasting roughly from 1972 to 1983, was characterized on the US side by introduction of smaller strategic nuclear weapons systems with improved accuracy - hence higher kill capability against hardened targets. Delivery systems were equipped with multiple, independently targetable reentry vehicles (MIRVs). On the Soviet side total weapons were proliferated, and the US lead was followed in areas such as MIRVing. The Soviet Union took the lead in the development of mobile land-based missiles (the SS-20 and SS-25).

By the early 1980s the US and the Soviet Union had roughly the same number of strategic weapons - about 10,000 on each side - though the mix and quality of technologies differed considerably. The arsenals of the United States and the Soviet Union have evolved both quantitatively and qualitatively over these years. In Figure 1 we show how the number of warheads in the strategic forces of the US and the Soviet Union have grown since 1945. In Figure 2 we show a similar plot for the number of launchers in the strategic arsenals of the two superpowers.

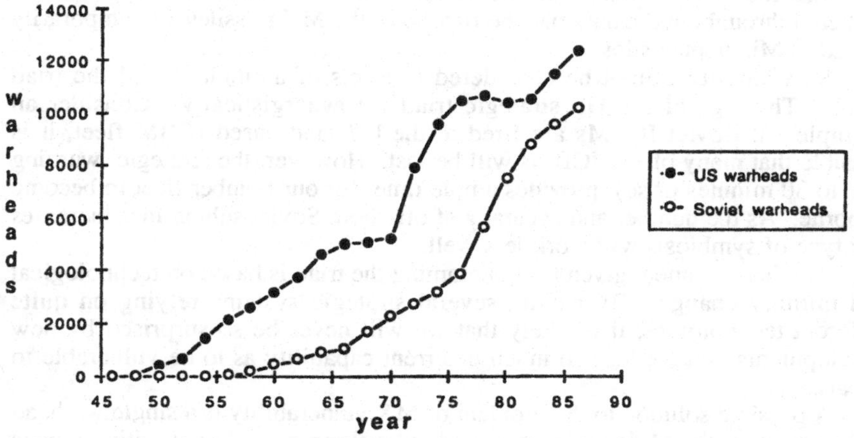

Figure 1. The number of warheads in the strategic forces of the US and the Soviet Union, as a function of time. Source: Robert Norris, William Arkin, and Thomas Cochran, *US-USSR Strategic Offensive Nuclear Forces 1946-1986*, Nuclear Weapons Databook Working Paper No. NWD 87-1, Natural Resources Defense Council, Washington, D.C.

Another way to view the evolution of the US strategic arsenal is in terms of the total megatonnage. This reached a peak of about 19,000 Mt (megatons) in 1959, and has since dropped (though not monotonically) to the present level of about 5000 Mt. The largest reduction was the result of dismantling in 1961 of the enormous weapons deployed on B36 bombers. Later reductions resulted from MIRVing of ICBMs and SLBMs. MIRVing results in an increase in warheads per delivery vehicle, but a reduction in total megatonnage on each vehicle. MIRVing occurred at a time when accuracy was increasing rapidly (2). This permitted a reduction in total megatonnage, with no sacrifice in lethality.

Both sides distributed the launchers among three different types of platforms: the land-based ICBMs, the submarine-based SLBMs and the bomber force. Figure 3 shows the distribution of US and Soviet forces among the three legs of their respective triads. . One notices the different distributions for the two countries, especially regarding the land-based ICBMs.

Technological developments on the Soviet side stimulated concern that the land-based leg of the US triad might be becoming vulnerable. A new MIRVed missile, the MX (later termed the Peacekeeper), was developed to have the accuracy to destroy hardened Soviet land-based ICBMs. To meet survivability criteria the missile would have to be sited so as to be invulnerable to a first strike. As Soviet accuracy continues to improve, numerous proposals have been made for basing the MX - ranging from deployment in extra hard silos to continual movement among shelters, to location on railroad cars which will be scattered throughout the nation. The first 50 of the MX missiles are temporarily placed in Minuteman silos.

Survivability cannot be considered in terms of a single leg of the triad alone. The legs of the US strategic triad act synergistically. Consider an example. If Soviet ICBMs are fired at the US land-based ICBM fleet, it is possible that many of our ICBMs will be lost. However, the strategic warning (up to 30 minutes or so) provides ample time for our bomber fleet to become airborne. As the number and accuracy of off-shore Soviet submarines increases this type of symbiosis will work less well.

Another argument given for maintaining the triad is based on technological and military change. By having several strategic systems relying on quite different technologies, it is likely that we will never be so surprised by new developments that we lose so much deterrent capability as to be vulnerable to threats.

A possible solution to the problem of MX vulnerability is a single warhead mobile missile, the Midgetman, now under development. Various siting modes for both Midgetman and MX have been discussed, all of which are designed to make them difficult to attack (e.g. super hardened silo, basing in or transporting on railroad cards or hardened mobile launchers, etc). We examine these and other options in this book.

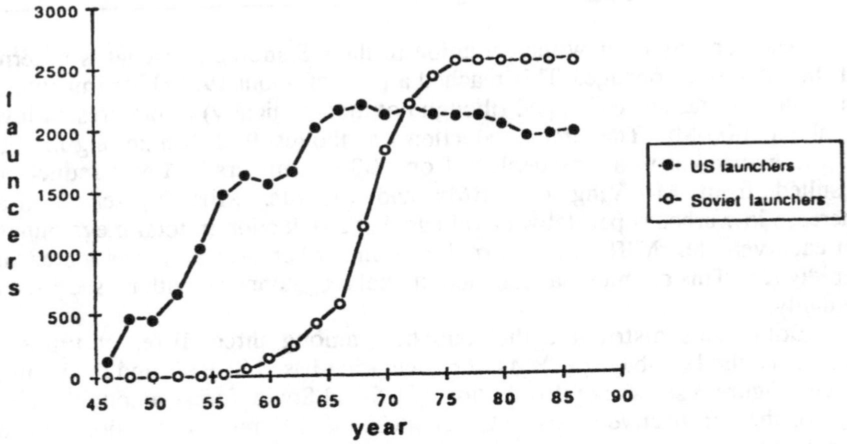

Figure 2. The number of launchers in the strategic forces of the US and the Soviet Union. Source: same as Figure 1.

In contrast to the United States triad, the Soviet strategic system more heavily emphasizes land-based ICBMs. This Soviet strategy is viewed by some United States analysts as unfortunate. According to this view we would like the Soviet Union ICBM fleet to be as invulnerable as our own. The recent Soviet moves toward mobile land-based missiles (the SS-20 and SS-25) are thus seen as stabilizing. These systems might be able to hide on land in much the way that submarines hide at sea.

The present ferment over the future of the land-based leg of the US strategic nuclear arsenal makes the next few years especially important. Decisions will be made which may affect fundamentally the nature of our national defense. They may also determine the feasibility and direction for future arms reductions. The issues are complex. They include military, technical, political, economic, social and perceptual issues (3).

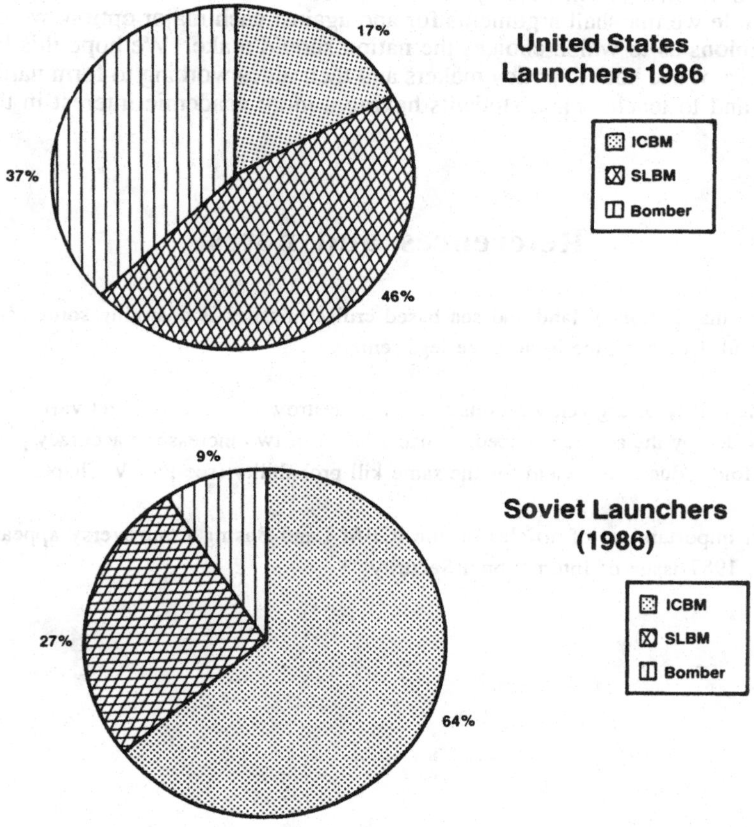

Figure 3. The distribution of launchers between the three legs of the strategic triad for the United States and the Soviet Union.

There are no easy answers to these questions. They are important to raise, and to analyze in depth. In this book we emphasize primarily technical issues - our area of competence. Other aspects are discussed when appropriate, but are treated in a much less comprehensive fashion than are technical matters. Our goal is to provide a review of the key facts and uncertainties which should be taken into account in national decision making. We have provided extensive references which will guide the reader to more extensive discussions of technical matters, and to the general literature in relevant areas which lie beyond the scope of this study.

We see three audiences for our work: the policy community interested in a balanced technical discussion at an accessible level, concerned citizens who want a deeper understanding of a complex and important issue, and the academic community concerned with the teaching of arms race issues. To meet the needs of these diverse audiences we have divided our report into a comprehensive forward section, supplemented by individual technical chapters.

While we marshall arguments for and against each major option, we offer few opinions as to which choices the nation should make. We hope this book will prove of use both to policy makers and their aides working to form national policy, and to teachers and students having a more academic interest in these matters.

References and notes

1. The installation of land and sea based cruise missiles is changing some aspects of the triad. However, the basic three legs remain.

2. The ability of a given weapons system to destroy a hardened target varies as the yield divided by the accuracy cubed. Hence a factor of two increase in accuracy permits an eightfold reduction in yield for the same kill probability (see Part V Chapter 3).

3. An important set of articles on the ICBM Land-Basing Controversy appears in the Fall, 1987 issue of International Security.

Part III

Nuclear weapons and strategic doctrine

Leo Sartori

The United States faces some critical decisions concerning the future of its strategic forces, particularly that of the land-based component. The ICBM leg of the strategic triad is widely perceived to be in need of modernization and strengthening; the debate is largely over what form that modernization ought to take. The present study analyzes the pros and cons of several options under consideration, including the option of doing nothing.

A rational discussion of the issues must begin by addressing several underlying questions:

•What is the mission of nuclear weapons in US defense policy?

•What do ICBMs contribute to the fulfillment of that mission? Is that contribution indispensable?

•What are the deficiencies in the present US force that require correction? Only with a clear definition of the problem can one assess the relative merits of the proposed solutions.

The role of nuclear weapons in defense policy

Deterrence. Virtually no one nowadays believes that the United States could escape vast destruction and an immense number of casualties in any conflict with the Soviet Union in which strategic nuclear weapons were employed on a large scale (1). The primary objective of US defense policy must therefore be to avoid such a conflict while protecting vital US interests around the world.

Discussion of the mission of nuclear weapons invariably focuses on deterrence-- the attempt to dissuade the Soviets from taking aggressive actions against the United States or its allies, through the threat of overwhelming nuclear retaliation. This definition is incomplete in two respects: it does not specify what Soviet actions we seek to deter, nor does it spell out what form the threatened retaliation would take. There is a divergence of views in the United States regarding both these questions.

In the most elemental definition of deterrence, the only action proscribed is a direct nuclear attack against the United States. Other types of deterrence can be defined by enlarging the set of provocations to which a nuclear retaliation is threatened. The expanded set can include nuclear attacks on US allies or even conventional attacks on the United States or on an allied nation; the expanded set is sometimes referred to as extended deterrence. US refusal to rule out first use of nuclear weapons in response to a conventional attack against NATO is an example of extended deterrence.

A critical element of any deterrent strategy is its credibility: if the would-be aggressor believes the threatened retaliation is not likely to be carried out, he may not be deterred. Inasmuch as nuclear retaliation to any aggression entails grave risks for the retaliator, credibility is far from automatic. In fact, as many authors have pointed out, deterrence theory suffers from a logical problem: if deterrence fails and an attack takes place, the victim might actually be worse off by carrying out the threatened retaliation than by not responding. In other words, the victim is himself subject to deterrence. Recognizing this, why should the aggressor be deterred in the first place?

Fortunately, successful deterrence does not require that the potential aggressor be certain his act would elicit the threatened response. Because the consequences are so grave, even a small risk of initiating large-scale nuclear war can be expected to deter under all but the most extreme circumstances. McGeorge Bundy has termed this *existential deterrence:* "As long as each side has thermonuclear weapons that could be used against the opponent, even after the strongest possible preemptive attack, existential deterrence is strong and it rests on uncertainty about what could happen" (2). In this sense all deterrence can be regarded as a bluff, but one that the adversary can call only at grave peril.

It stands to reason that the credibility of a deterrent threat diminishes as the set of provocations against which it is directed expands. A prudent Soviet planner must assume that a nuclear attack on the United States would in all likelihood elicit a nuclear response, regardless of the consequences. But is the United States really prepared to risk all-out nuclear war in response to a lesser provocation, say a conventional attack on one of the secondary NATO countries? That threat must be considered less credible.

With deterrence as the guiding policy principle, the fundamental problem of strategic planning is to determine what kind of force and what targeting plan will provide the most effective and most credible deterrent. The solution to the problem depends on what view of deterrence one adopts. If one is willing to rely on existential deterrence, for example, the detailed nature and size of the US nuclear force become almost irrelevant. About the only requirement is that at least part of the force be able to survive a preemptive attack.

In practice, existential deterrence has never had much influence on US strategic policy. It has been criticized as being a non-strategy, in the sense that it provides no guidance whatever as to what the United States ought to do in case deterrence fails and we are attacked. This, in the eyes of some, makes it useless for planning purposes.

The school of thought that has dominated strategic planning for some time holds that for credible deterrence the United States must have the capability to retaliate to nuclear attack with something other than a spasmodic strike directed against Soviet civilian/ industrial targets (countervalue second strike), and must have a variety of options available to match the variety of possible attacks. Terms such as "flexible response" are used to describe such an approach.

Some go further and argue that the United States must be equipped to wage a limited though possibly protracted nuclear war. They contend that if properly prepared, the United States can limit the damage it suffers, achieve its major political objectives, and emerge from such a conflict with a functioning society that can quickly recover. Such a "war-fighting" doctrine clearly goes beyond pure deterrence, although its proponents argue that the capability to wage nuclear war "successfully" is itself the most effective deterrent. Others strongly disagree, contending that the quest for a war-fighting capability leads to dangerous instabilities and could indeed make war more likely. We shall not discuss these doctrinal issues further here, but note that they have very significant implications for both the character and the size of the nuclear force that is required.

Counterforce. The preceding considerations lead logically to a discussion of counterforce, the capacity of a weapon to destroy hardened targets such as missile silos. Counterforce is arguably the most controversial characteristic of nuclear forces. Some would say that counterforce has no part in a pure deterrent strategy because a retaliatory strike would be aimed at countervalue targets which are not hardened. The contrary position is based on the assumption that any Soviet first strike would be a counterforce attack designed to knock out US strategic forces while minimizing collateral damage to civilian population and industry. A countervalue retaliation to such an attack would be suicidal, so the argument goes, because it would inevitably elicit a devastating second-wave Soviet attack against U.S. cities and would do nothing to blunt the force of such an attack. Inasmuch as a threat to commit suicide in response to any attack has low credibility, a strategic force possessing only countervalue capability would have little deterrent value.

With counterforce capability, on the other hand, the United States would have the allegedly more credible option of responding with a counterattack confined to Soviet military/ strategic targets-- a counterforce second strike-- which is less likely to trigger a disastrous exchange of attacks on population centers. From this point of view, counterforce capability is an essential component of a credible deterrent force (3).

But the credibility of a purely counterforce strategy has been questioned by studies of the civilian casualties resulting even if only strategic nuclear forces are attacked. Tens of millions may die, blurring the distinction between countervalue and counter force (4).

The preceding argument overlooks the fact that most military targets, including such strategic installations as airfields and submarine bases, are relatively soft. Hence a countermilitary response to an attack does not necessarily require counterforce capability. As a matter of fact, although hardened missile silos have the highest strategic value, most of the targets in the SIOP have always been and continue to be soft.

The distinction between counterforce and countervalue targeting is, in this light, somewhat misleading. A better approach (5) would be to classify targets into three groups:

(i) Strategic weapons and support facilities;

(ii) Other military targets, including both nuclear and conventional military infrastructure; and

(iii) Industrial targets (oil refineries, power plants, and so on) vital to a continuing military effort. Only targets in group (i), and not all of those, require counterforce capability; as May et al point out (5), group (ii) comprises the targets most appropriate to a deterrent strategy.

Another criticism of the argument for counterforce is that a retaliatory strike aimed at hardened silos is likely to hit mostly empty holes, since the missiles that filled those holes would have been launched as part of the postulated first strike. Any enemy missiles that had been withheld from the first strike would surely be launched on warning if attacked; there would be no question of a false alarm under those circumstances.

The paradox posed by counterforce is that any force capable of carrying out an effective counterforce second strike can *ipso facto* carry out an even more effective counterforce *first* strike. Acquisition of a powerful counterforce capability by either superpower is therefore bound to be viewed by the other as an intolerable threat to its deterrent, one that requires a response. Even a trend in that direction is likely to cause alarm. The growth of Soviet counterforce capability during the 1970s was in fact so viewed by the United States, giving rise to the "window of vulnerability" argument which provided justification for substantial expansion of US strategic forces.

Each side tends to view the other's quest for counterforce capability as an indication of aggressive intent while proclaiming the purely defensive character of its own counterforce. Possession by both sides of strong counterforce capability is destabilizing in that it provides an incentive to strike first in a crisis.

In sum, the question whether counterforce capability is necessary for successful deterrence remains controversial. This question plays an important part in the current debate over deployment options. Doctrinal considerations aside, it is a fact that US counterforce capability has been increasing over the past decade, and that most proposed new deployments would continue that trend.

Attributes of nuclear weapons relevant to a deterrent strategy

In this section we briefly describe the principal attributes of strategic weapons that determine their capability to carry out the missions of a deterrent strategy, and that bear on the evaluation of competitive deployment options. As noted above, there is a considerable divergence of views as to what such a strategy entails.

Survivability. This is the single most essential attribute for a second-strike weapon. If a weapon system is vulnerable to being destroyed in a pre-emptive attack its deterrent value is slight, since it cannot be relied upon to be available for the threatened retaliation.

It is generally assumed that the United States would choose to ride out any Soviet first strike, at least to the point at which nuclear weapons had detonated on US soil and the scale of the attack had been unambiguously determined. Under a policy of launch on warning (LOW) or launch under attack (LUA), survivability is not a critical consideration; more important requirements in that case would be a reliable early warning system and the ability to launch weapons quickly. But LOW (and to some extent LUA) are generally held to be reckless policies because they risk initiating nuclear war as the result of an accidental launch or even of a false alarm. Moreover, because of the extremely short time available, they may require entrusting crucial launch decisions to a computer (6).

If LOW is rejected, survivability becomes a primary criterion in evaluating any weapon system; the sole motivation for all mobile missile basing proposals is to improve survivability. Extensive deployment of vulnerable systems could be construed as evidence of aggressive intent, since a first striker need not be concerned with the survivability of his forces. Emphasis on survivability, on the other hand, is an indication that first-strike options are not being entertained.

Accuracy and lethality. These are the main characteristics that determine counterforce capability. In order to produce the great overpressures needed to destroy a highly hardened target, a missile must land very close to the target. The accuracy required depends on the hardness of the target and, to a lesser extent, on the nuclear yield of the weapon. Lethality is a parameter that depends on both accuracy and yield and that measures a weapon's ability to destroy hardened targets. (See Part V Chapter 3 for a precise definition.)

Time to target. This is another attribute associated with counterforce. A very slow weapon, even if it has high lethality, gives the opponent time to implement countermeasures, including possible launch of the missiles under attack (7). Hence an effective counterforce capability requires that at least part of the force be capable of reaching its targets quickly.

Security and reliability of command, control, and communications (C3). These are desirable characteristics for strategic weapons systems in any case, but particularly for a second strike mission. If the United States is to ride out an attack and retaliate effectively, the systems involved must be able to receive commands and respond promptly and reliably, and to do so in the highly stressed environment that would follow the detonation of large numbers of nuclear weapons over US territory. This requirement calls for a highly robust C3 system, including back-up modes for vulnerable components. The requirement is more stringent for war fighting scenarios where intended targets may change then for countervalue strikes where missile destinations are preprogrammed. A first striker, on the other hand, has ample time to plan his attack. His weapons would be programmed on their targets perhaps long before any crisis precipitated their launching.

Arms control compatibility. Since no treaty limiting strategic systems is now in effect, each side is free to deploy new weapons of whatever type and in whatever quantity it pleases. However, negotiations on a strategic agreement (START) are ongoing and considerable progress has been made; each side has proposed ceilings that would require substantial reductions from present force levels.

The possibility of a START agreement influences deployment decisions in two ways. First, whatever weapons are selected for deployment should be ones that the United States would plan to retain as part of its reduced force; there is no point in deploying expensive new weapons only to have to dismantle them if a treaty comes into effect, except possibly for use as bargaining chips in the negotiation. Second, the implications of any proposed deployment for the verifiability of a START treaty must be taken into consideration.

At the heart of any treaty would be ceilings on deployed missiles and/or warheads; the basic verification task would therefore be to count those deployments. Weapons that are difficult to count would complicate that task. Verifiability must be part of the assessment of the relative merits of proposed deployment modes for land-based missiles. In general, fixed (silo-based) missiles are easy to count while mobile missiles pose a greater challenge. Each mobile mode presents its own unique verification problems, which are addressed in detail in Part IV (Chapters 5 and 6) and Part V (Chapters 6, 7, 8, and 10).

Crisis stability. Many people feel that the most plausible failure mode for deterrence is through miscalculation or panic in a crisis situation. Leaders who would not contemplate initiating a nuclear attack in a period of calm might do so under the pressure of a crisis if convinced that the other side was about to attack and that a relative advantage could be gained by striking first. Crisis stability refers to a situation in which no advantage, real or even perceived, accrues to the side that gets in the first blow; under such conditions a panic-driven attack is much less likely.

A crude measure of stability is the *exchange ratio*, defined as the number of warheads destroyed per weapon employed in an assumed first strike. The higher the exchange ratio, the more unstable the state.

Crisis stability is closely correlated with weapon survivability. Vulnerable forces make tempting targets in a crisis. At the same time their possessor, fearful that they will be destroyed in a first strike, may be tempted to launch them first. (This is sometimes called the "use them or lose them" argument.) If forces are highly survivable, on the other hand, there is no incentive either to attack them or to launch them preemptively in a crisis. Such a state is stable.

Missiles with MIRVs are often cited as a major source of crisis instability because they create the possibility of destroying several weapons with a single attacking warhead. The problem is particularly acute if highly MIRVed missiles are based in a mode that has low survivability. The exchange ratio in this case can become quite high.

Although individual weapons can be characterized as stabilizing or destrabilizing, it is important to keep in mind that crisis stability is a property associated with the *totality* of weapons on both sides. In a crisis, each side must compare the likely costs of launching or absorbing a first strike. The capability to destroy one component of the adversary's force will not necessarily make a first strike attractive if the components that survive can inflict sufficient damage. Such a state must be judged to be stable. This is in essence the situation at the present time.

Costs. The cost of a weapon system does not directly affect its strategic utility. As a practical matter, however, cost considerations can be expected to play a significant role in any deployment decision. The prospects for Midgetman, which is attractive on many other grounds, are diminished by its high unit cost. The operative criterion in the end is likely to be how much capability can be bought for each dollar expended.

Capabilities of land based missiles

Subsequent chapters will assess in detail how each of the proposed deployments rates in terms of the attributes described above. Here we make some general remarks concerning ICBMs and compare their capabilities to those of other strategic weapons.

Land-based missiles have been a pillar of the US strategic force since the deployment of Minuteman I in the early 1960s. Their properties today can be summarized as follows.

•Survivability: low in silo basing, as the result of improvements in the accuracy of Soviet missiles. This is the major perceived weakness of present-day US ICBMs and the principal reason why modernization is felt to be necessary. Several different ways of improving survivability have been proposed.

•Accuracy/lethality: high. Minuteman III has a CEP of 220 meters, while MX is the most accurate intercontinental missile yet deployed by either the United States or the Soviet Union (CEP estimated at about 90 m.) With 10 warheads of yield estimated at 335 kT, MX provides substantial counterforce capability. A force of 100 MX, the deployment proposed by the Reagan Administration, would contain 1000 warheads, not enough to threaten the existing Soviet ICBM force of about 1400 launchers; however, in combination with other counterforce missiles it could do so. It should be noted in this connection that the proposed Trident D-5 SLBM is expected to have accuracy and lethality that nearly match those of MX. Midgetman will also have high accuracy; with only a single RV, however, its lethality per missile is substantially less than that of MX.

•Time on target: fast. The only shorter transit time would be for an SLBM launched from a point close to the coast of the USSR.

•C3: fixed ICBMs are considered to have highly secure communications, although the reliability of the C3 system is questioned from time to time. Communicating with mobile missiles is an intrinsically more complicated task; this is one of the major problems being studied in connection with proposed mobile basing modes.

•Crisis stability depends on the type of missiles deployed as well as on their basing modes. A highly MIRVed system with high counterforce capability, based in vulnerable silos, is the epitome of instability. It can be a potent force if launched first but is itself liable to be destroyed in a first strike by a numerically smaller attacking force. Mobile basing improves stability by reducing vulnerability. A single RV missile like Midgetman promotes stability because it is a relatively unattractive target. The exchange ratio in attacking an unMIRVed missile cannot be greater than unity unless the missiles are based so close together that more than one can be destroyed by a single attacking weapon.

Each of the other two legs of the strategic triad, SLBMs and bombers, has some advantages and some disadvantages relative to ICBMs. SLBMs are more survivable but have less reliable communications. They are at present considerably less accurate then ICBMs but that situation will change when Trident D-5 enters the force.

Bombers are much slower than ICBMs and are subject to extensive Soviet air defenses. (Stealth technology is being developed in order to improve bomber penetrability.) They are survivable when airborne but highly vulnerable on the ground. They offer the advantages of being recallable and readily retargetable. The synergistic relation between bombers and ICBMs, which makes simultaneous surprise attacks on both impossible, is discussed elsewhere in this study.

The unique characteristics of ICBMs make them the weapon of choice for attacking critical time-urgent targets, which include (but are not restricted to) hardened sites. The prevailing view in Washington is that the United States must have the capability to strike such targets and therefore that it is essential to repair the deficiencies in the present ICBM force. Most of the options analyzed in our study represent different proposed ways of achieving that objective. A minority view holds that the overall US strategic force in its present state provides a robust deterrent and that the threat to US security due to the ICBM vulnerability problem has been exaggerated. From that point of view, the option of leaving the ICBM force essentially unchanged is one that merits serious consideration.

It is important to bear in mind the distinction between the two principal motives for the proposed modernization of the ICBM force-- to remedy the vulnerability of silo-based ICBMs and to provide increased counterforce capability. Any assessment of the relative merits of proposed solutions depends strongly on which motive is given greater emphasis. If survivability is the principal objective, the basing mode of whatever missile is deployed is much more important than the characteristics of the missile itself. Deploying Minuteman III in any of the proposed mobile modes would achieve the same survivability. If, on the other hand, counterforce is considered the main objective, the relative rankings are very different.

Negotiating history

Mobile missiles have had a checkered history at the negotiating table; both sides have reversed their positions more than once on the issue, guided by the state of their own programs as well as of those of the adversary.

During the SALT I negotiation, the Nixon Administration proposed to ban deployment of all mobile ICBMs. That proposal was not withdrawn until the very end, when it became clear that the Soviets would not agree to it. In a unilateral statement issued on the eve of the Treaty signing in 1972, Ambassador Gerard Smith said that the United States "would consider the deployment of operational land-mobile launchers during the period of the Interim Agreement as inconsistent with the objectives of that Agreement." The basis for the claimed inconsistency was never spelled out. In any event, neither side deployed any mobiles during the period in question.

As SALT II was being negotiated, the problem of Minuteman vulnerability was at the center of attention in the United States. The Carter Administration proposed to address the problem by deploying the new MX missile in a mobile mode. Numerous possible basing modes were studied; the one finally selected was the multiple protective shelter or "shell game" scheme. The United States negotiating position at SALT was tailored to protect the option of proceeding with that deployment (8).

The Soviets apparently sought a ban on mobiles in SALT II, (9), although they were themselves then flight testing a new mobile ICBM, the SS-16. Unable to agree on a permanent solution, the sides decided as a temporary measure to ban flight testing and deployment of mobiles in the SALT II Protocol (10). Inasmuch as the Protocol was scheduled to expire on December 31 1981, long before MX was due to be flight tested, the provision had no operational impact on US programs. It likewise had no direct effect on Soviet programs because the Soviets had already agreed not to produce, test, or deploy the SS-16.

Only two explicit references to mobiles appear in the SALT II Treaty proper. One is a ban on deployment of mobile launchers of heavy ICBMs, something that neither side was likely to contemplate doing in any case. The other is a statement that after the expiration of the Protocol, mobile launchers would be subject to all the limitations applicable to ICBM launchers, "unless the parties agree that mobile ICBM launchers shall not be deployed after that date."

Limitations on mobiles were expected to be one of the major issues to be addressed in SALT III, which was to follow immediately upon the ratification of SALT II. The Joint Statement of Principles for Subsequent Negotiations, which accompanied SALT II, cites the need for "cooperative measures contributing to the effectiveness of verification by NTM". The reference to cooperative measures was included at the insistence of US negotiators, in the expectation that SALT III would permit mobiles and that verification of mobile deployments would be one of the tasks for which cooperative measures would be required. The Soviets insisted on explicitly tying cooperative measures to NTM so as to avoid what might be interpreted as an open-ended commitment that could lead to intrusive verification measures.

In the START negotiations, the positions of the two sides were once again reversed. The Reagan Administration proposed to ban all mobiles even as it actively pursued possible mobile deployments for MX and/or the new SICBM. A significant change in the situation since the time of SALT II is that Soviet mobiles have in the meantime become a reality. Deployment of the road-mobile SS-25 began in 1985 and the MIRVed SS-24 is now being deployed in a rail-mobile mode.

The proposal to ban mobiles has elicited no interest on the part of the Soviets. Their lack of enthusiasm can hardly be considered surprising. With US deployment of MX already begun and that of D-5 SLBMs possessing hard-target kill capability looming on the horizon, the Soviet military leadership must be increasingly concerned over the potential threat to their silo-based ICBMs. Inasmuch as the preponderance of the Soviet strategic capability resides in land-based missiles, the potential vulnerability of silo-based ICBMs represents an even more serious problem for them than it does for the United States. As Soviet mobile deployments proceed, a ban on mobiles will become even less attractive to them, since it would require them to dismantle or redesign an ever increasing fraction of their most modern missiles.

It seems increasingly unlikely, then, that a ban on mobiles will be included in any future agreement. If the United States decides to proceed with mobile basing of either MX or Midgetman, the proposal to ban mobiles will presumably be withdrawn. This would advance the prospects for a START agreement by removing one of the major issues that divide the two sides.

Why has the Reagan Administration been so anxious to ban mobiles? Concerns over verification are the most commonly cited reason. Although those concerns are legitimate, other possible motives suggest themselves. One is the substantial Soviet head start in mobile deployments: by the time either of the proposed US systems is ready to be deployed, the Soviets could have hundreds of mobiles in the field. And if no mobile basing plan is approved and funded by the Congress, the Soviet advantage would be perpetuated.

Another relevant consideration is that geography confers definite advantages on the USSR with respect to mobile basing. Vast sparsely populated areas in Siberia offer attractive opportunities for mobile deployments. Preservation of location uncertainty, vital for survivability, could be more reliably achieved by the Soviets than by the United States. Moreover, concern for American public opinion rules out many basing schemes that might be technically attractive. For example, any proposal that calls for missile-carrying trains to travel routinely over the public rail system in peacetime would generate intense public protest and would almost surely be rejected by the Congress. Recall that civilian protests over environmental and land-use issues doomed the Carter Administration's proposed MPS basing scheme. The Soviet leadership does not have to worry about such concerns.

A final possible motive for wanting to ban mobiles is that the Administration simply does not want the Soviets to have a highly survivable land-based force. The existence of mobile Soviet ICBMs no doubt complicates the Pentagon's targeting plans (11). The underlying question is whether US security is better served if both sides have invulnerable land-based forces, or if neither one does.

Constraints created by existing agreements

In May 1986 President Reagan announced that the United States would no longer be bound by the informal commitment not to "undercut" existing agreements (i.e., the unratified SALT II and the expired SALT I Interim Agreement.) As a consequence of that decision, no limitations on strategic offensive arms are currently in force. Nonetheless, there remains considerable sentiment within the Congress in favor of returning to compliance with SALT; both houses have voted on several occasions to do so. It is noteworthy that the Soviets have continued to dismantle SALT-accountable systems as they deploy new ones, so as to stay within the SALT ceilings, more than two years after the President announced that the treaty was dead. For the present at least, they are apparently leaving the door open to the United States to return to compliance. A new US administration might well decide that SALT II provides a useful set of interim restraints, pending completion of a new START treaty. Such a decision would have a number of programmatic implications for the deployment options under consideration, none of them very serious.

Mobile missiles and arms control

Land-mobile missiles constitute a dilemma for arms control. The sole purpose of mobile basing, to enhance survivability, is fully compatible with the objectives of arms control. In a strategy based on deterrence, the fact that the other side's forces are survivable should not be a concern. With increased confidence in the survivability of their own forces, moreover, both superpowers ought to be more willing to accept substantial reductions in the numbers of deployed weapons.

On the other hand, mobiles pose a potentially serious hazard for arms control because they complicate the task of verifying compliance with treaty limitations. Concealment, the very feature that is relied upon to provide survivability, necessarily complicates the task of counting mobile missiles. If the numbers of deployed missiles and/or warheads cannot be determined with high confidence, any treaty provision that sets a ceiling on those numbers becomes unverifiable and therefore unacceptable, particularly to the United States which has always emphasized verifiability as an absolute requirement.

The problem is, then, how to make mobile deployments verifiable without compromising the location uncertainty from which survivability stems. This problem has been under study for more than a decade in the United States, principally in connection with proposed mobile basing modes for the MX missile (12).

That the verification problem is not insoluble is suggested by long experience with submarine basing, which likewise provides survivability through concealment. Both the United States and the USSR maintain reliable counts of each other's deployed SLBMs, not by being able to see all of them at any given time but by monitoring the construction of missile- carrying submarines and observing them as they go to sea and periodically return to port for servicing. National technical means of verification (NTM) are entirely adequate for these tasks.

Unfortunately, many of the characteristics that facilitate counting deployed SLBMs are not shared by land-mobile missiles. Submarines are constructed over a period of several years at readily identifiable facilities. They are based at only a few ports, from which their departure can be unmistakably detected by satellite photography. None of this is true of land-mobile ICBMs. It has long been recognized that measures going well beyond NTM would likely be required if limitations on mobile ICBM deployments were to be made verifiable. The term "cooperative measures" has come to be used to describe agreed production and deployment practices specifically designed to assist in verification.

Several types of cooperative measures have been suggested that could contribute to verification of START provisions dealing with mobile ICBMs. They include:

•Measures to facilitate counting the number of missiles produced. This might require routine stationing of inspectors at plants where missile components are manufactured or assembled. Proposed schemes to "tag" missiles at production can be included under this category;

•Rules that restrict deployment to designated geographic areas, preferably with limited access routes to facilitate monitoring the movement of missiles into and out of the deployment area. Detection of even one operational missile outside the designated areas would then constitute *prima facie* proof of a violation;

•Sampling techniques, applicable to deployments in which missiles are shuttled among a large but finite number of well-defined locations; and

•Challenge inspection in case of a suspected violation.

Many of these measures were until recently regarded as totally out of reach because of long-standing Soviet objections to any kind of "intrusive" verification. In particular, any proposal that called for on-site inspection was considered certain to be rejected by the Soviets. The Soviet attitude toward verification appears to have changed dramatically, however, with Mikhail Gorbachev's accession to power in the Soviet Union. In the recently negotiated INF Treaty the Soviets agreed to an extensive set of verification measures, including some on-site inspection. Inasmuch as the INF missiles are themselves mobile, at least some of the procedures adopted in the treaty are relevant to the problem of verifying treaty provisions that govern mobile ICBMs.

The INF experience does not, however, guarantee that compliance with limitations on mobiles in a START treaty can be verified successfully. Verification of the INF treaty is simplified by the fact that the missiles in question are totally banned; hence detection of even one missile would be sufficient to establish a violation. Verifying compliance with treaty provisions that permit mobile ICBMs but set numerical ceilings on the number deployed is a much more challenging task.

Useful lessons have been learned from US experience in monitoring Soviet mobile deployments-- the SS-20 IRBM and the SS-25 ICBM-- over the past ten years or so. Even though neither deployment has been restricted in any way by treaty obligations, US intelligence has apparently been able to estimate with fairly high confidence the numbers of missiles deployed. These estimates are based primarily on careful observation of Soviet operational procedures. Typically, a squadron of missiles is housed in a central support facility, from which the missiles disperse to (unknown) outlying locations. The number of missiles associated with each support facility can be inferred from the size of the buildings and other similar data.

Estimates of missile counts obtained in this manner are obviously not absolutely reliable. In principle, there might be hundreds of additional SS-20s roaming unobserved through the woods, unattached to any support facility. That seems most implausible, however, because nuclear missiles are complicated devices that require sophisticated equipment and trained personnel for maintenance, as well as elaborate security arrangements. Moreover, transporting large numbers of missiles surreptitiously to their deployment sites would be a difficult and risky undertaking.

Still, there is no doubt that if the side carrying out a mobile deployment wanted to make it difficult for the other side to count the number of missiles deployed, it could find ways to do so. Numerous evasion scenarios can be imagined. Hence in a treaty regime a set of operating procedures would have to be spelled out and made mandatory. The details would depend on the nature of the basing mode and would have to be worked out on a case-by-case basis. A joint commission like the SCC could be assigned responsibility for this task.

Additional discussion of the verification problem is found in Part V Chapter 2.

References and notes

1. If SDI were to be successfully developed and deployed, this assessment could change. We express no opinion on the feasibility of SDI, but simply note that even according to the most optimistic estimates, no effective area defense could be deployed until well into the next century. Hence the prospect of strategic defense can have little influence on near- or intermediate-term strategic planning.

2. McGeorge Bundy, "The Bishops and the Bomb," NY Review of Books, 16 June 1983.

3. Another argument for counterforce is that because the Soviets have the capability, the United States must have it as well in order to maintain strategic parity.

4. William H. Daugherty, Barbara G. Levi, and Frank N. von Hippel, "The consequences of 'limited' nuclear attacks on the United States," International Security, Spring 1986, pp. 3-45.

5. M. M. May, G. F. Bing, and J.D. Steinbruner, "Strategic arsenals after START: the implications of deep cuts," International Security, Summer 1988, pp. 90-133.

6. It is worth noting that a potential aggressor can never be certain the victim's missiles will not be launched on warning, even if that is not his declaratory policy. This possibility increases the uncertainty in any estimate of the likelihood of success of a pre-emptive strike and may be regarded as an element of existential deterrence.

7. Provided, of course, that the adversary is able to detect and identify the weapon. In the case of cruise missiles, this is not a trivial task.

8. There was some difference of opinion in US circles as to whether MPS basing was compatible with the SALT prohibition on construction of new fixed launchers of ICBMs. The dispute hinged on the technical question whether the individual shelters should be considered launchers. The Pentagon took the position that they should not, because all the equipment necessary to launch would travel with the cannisterized missile. Critics disputed this claim. Paul Nitze argued that at least the vertical-shelter version of MPS would be barred; this was one of his principal reasons for opposing the Treaty.

9. S. Talbott, *Endgame: the Inside Story of SALT II*, Harper and Row, New York (1979).

10. The protocol also banned flight testing and deployment of air-to-surface ballistic missiles, which can be considered another type of mobile ICBM.

11. It is reported that one of the primary missions for the new Stealth bomber will be to hunt down Soviet mobiles.

12. MX is not the first US ICBM for which mobile basing has been considered: Minuteman was initially designed to be deployed partly in rail basing. This was, however, in the pre-SALT era when verification was not a concern. The plan for mobile basing never got very far.

Part IV

Evaluation of options

*Paul Craig, David Hafemeister,
Art Hobson, Ruth H. Howes,
Barbara G. Levi, John Michener,
Mark Sakitt, Leo Sartori,
Valerie Thomas,
Peter D. Zimmerman*

Part IV Chapter 1

Evaluation: No upgrade of land-based missiles

If the US does not deploy new land-based missiles, it must rely on its silo-bound Minutemen and MX missiles. The US currently has 450 single-warhead Minuteman II missiles, 200 three-warhead Minuteman III missiles, and 300 three-warhead Minuteman IIIA missiles (nearly identical with Minuteman IIIs, but carrying twice the yield). It has deployed 50 MX missiles in Minuteman silos. Details of these missiles are given in Table 1. We evaluate here the different attributes of this current force for comparison with other options.

Survivability

Minuteman/MX (MM/MX) silos will soon be highly vulnerable to ICBM attack, and perhaps to SLBM attack. Under pessimistic "worst case" assumptions, a two wave ICBM attack against the 1000 US ICBM silos in the mid-1990s would destroy all but a handful (1%) of those silos, and even a single wave attack would leave "only" 100 (10%) surviving silos. Under equally plausible "best case" assumptions, 140 silos (14%) would survive the two-wave attack.

If the Soviets deploy an SLBM comparable to the US Trident II, US silos could in addition be highly vulnerable to SLBM attack. Thus the greater accuracy of SLBMs could threaten the "synergistic survivability" of US ICBMs and bombers. Today US ICBMs can be effectively attacked only by highly accurate Soviet ICBMs, whereas US bombers can be attacked only by short-warning offshore SLBMs, so that timing problems prevent the Soviets from destroying both the ICBMs and the bombers simultaneously. It is at least plausible that the Soviets will deploy an SLBM rivaling Trident II (to be deployed beginning in 1989). In this case US ICBMs and bombers will be simultaneously vulnerable. The SLBM attack would need to be large: some 16 offshore submarines (2000 warheads) targeted against US ICBMs, in addition to the submarines needed to target US bomber bases. If the Soviets routinely deployed such large silo-destroying forces off of US shores, the submarines would be the only clearly survivable US strategic force.

Since survivability of US weapons depends strongly on the Soviet nuclear arsenal, it is best studied within the context of specific assumptions about Soviet weapons. Thus we look at four scenarios: (1) an all-out arms race in which Soviet strategic forces are doubled by the mid-90s; (2) continuation of the

Table 1. Characteristics of the presently-deployed US ICBMs, and Midgetman.

	MM II	MM III	MM IIIA	MX	Midge
Weight (kg)	33,000	35,000	35,000	87,000	17,000
Length (m)	17.7	18.3	18.3	21.6	16.2
Diameter (m)	1.7	1.7	1.7	2.3	1.2
Range (km)	11,000	11,000	11,000	11,000	11,000
No. boost stgs	3	3	3	3[c]	3[c]
Propellant	solid	solid	solid	solid	solid
Throw-wt[a] (kg)	700	700-900	700-900	3400	450
warheads/missile	1	3	3	10	1
RV designation	Mark-11C	Mark-12	Mark-12A	Mark-21	Mark-21
Warhead desig.	W56	W62	W78	W87	W87
Yield (MT)	1.2	0.17	0.35	0.3	0.3-0.5
Guidance syst.	NS-17	NS-20	NS-20	AIRS	mod-AIRS
CEP[b] (m)	370	220	220	90	90-120
Silo strength (atm)	133-200	133-200	133-200	133-200	mobile
Total missiles (1989)	450	200	300	50	0
Total warheads (1989)	450	600	900	500	0

Notes

a. "Throw-weight" means the weight launched toward the target, i.e. the weight of all the re-entry vehicles plus the MIRV carrier (if the missile is MIRVed) plus any penetration aids.

b. The CEP is that distance, measured from the target, within which there is a 50% probability of the warhead hitting.

c. The MX and Midgetman carry a post-boost liquid propellent stage for final accuracy adjustment.

present SALT II limitations; (3) 50% cuts in strategic forces; and (4) a finite deterrence regime in which each side is allowed "only" 500 ICBM warheads and 500 SLBM warheads. Assuming no changes in US ICBMs, how plausible and how successful would a Soviet attack be in each scenario?

Some 2000 warheads are needed for a 2-wave attack on the MM/MX force. Under either arms control scenario (1) or (2) or (3) above, the Soviets would have plenty of missiles, either ICBMs or SLBMs, to accomplish this with hundreds of SLBMs left for a short-warning attack on US bombers. However an ICBM attack on the MM/MX force would give the bombers enough warning to escape, whereas an SLBM attack from offshore submarines could catch not only US ICBMs but also the bombers at or near their bases.

The MM/MX system will be reasonably survivable in the mid-1990s only if Soviet forces are subject to restrictions similar to the finite deterrence limits. But in this case the question of MM/MX survivability is moot because these missiles would have to be partly or entirely removed to conform with those same limits. In the spirit of our "no upgrade" option, the US could in this case remove all of its MIRVed ICBMs and retain only its 450 unMIRVed MM IIs. It would certainly be unwise to retain any vulnerable MIRVed ICBMs in such a strongly arms-controlled world, because Soviet attack with a small fraction of their warheads could then destroy a large fraction of ours. The Soviets could attack unMIRVed ICBMs also, of course, but they would inevitably use more warheads than they would destroy. The US would wind up relatively stronger after the exchange, a fact which should deter the attack in the first place.

Synergistic survivability of the US bomber and MM/MX forces could be maintained if Soviet SLBM accuracy were restricted via an arms control agreement. Significant upgrades in accuracy require flight testing, and such tests of new or modified SLBMs could be banned. Unfortunately, US testing and near-deployment of the highly accurate Trident II probably make this option obsolete.

Another helpful arms control measure would be a ban on the testing of SLBMs along depressed trajectories. Such a "fast SLBM" capability requires testing, but it has not yet been tested by either side (1). This limitation would give the US more confidence in the ability of its bombers to escape a short-warning SLBM attack.

Lethality

The 10-warhead MX missile is very "lethal" against hardened targets such as missile silos. Quantitatively, if a single MX warhead struck a typical silos strengthened to 200 atm the silo would have a "single-shot probability of survival" (SSPS) of only 2%. For comparison, if targeted by the 10-warhead Soviet SS-18, described by the Reagan Administration as the world's most destabilizing weapon, a typical silo would have a much larger 40% SSPS. The 3-warhead US Minuteman IIIA is also far less lethal than the MX, each warhead being about as lethal as an SS-18 warhead.

How does this lethality look from the point of view of Soviet worst case planners? In the worst case (for the Soviets) the 500 MX warheads would be launched in a single wave against 500 Soviet silos, destroying 440 of them (assuming 90% reliability), while the 900 Minuteman IIIA warheads would be launched in a two-wave attack against another 450 silos, destroying 360 more silos. These 950 attacked silos would include all of their highly MIRVed SS-18s and SS-19s (668 silos holding about 50% of the total Soviet strategic warhead inventory), plus nearly 300 other silos. The Soviets would retain some 600 ICBMs (out of about 1400) and 1600 ICBM warheads (out of about 6800). However, many of these may be threatened by the US Trident II SLBM scheduled for initial deployment in 1989.

Stability

The Scowcroft Commission, which in 1983 launched the current ICBM modernization, defined stability as "the condition which exists when no strategic power believes it can significantly improve its situation by attacking first in a crisis or when it does not feel compelled to launch its strategic weapons in order to avoid losing them" (2).

For any strategic force, the key stabilizing technical characteristics are: survivability, so that the other side does not "believe it can significantly improve its situation by attacking first in a crisis," and inability to attack the other side's deterrent forces, i.e. low lethality, so that the other side "does not feel compelled to launch its strategic weapons in order to avoid losing them."

The silo-based MM/MX force's low survivability and high lethality is a destabilizing combination. The MM/MX is further destabilized by the MIRVing of Minuteman III and MX, because a more highly MIRVed missile makes a more tempting target. A good measure of this feature is the so-called "attack price": the number of warheads needed for the attack divided by the number of warheads destroyed. (The inverse of this number, i.e. the number destroyed divided by the number used, is also known as the "warhead-exchange ratio.") See Part V, Chapter 1.

To all of this must be added the instability arising from the pressure to launch these vulnerable missiles on tactical warning, i.e. on receiving word that a Soviet attack is underway. Obviously, weapons that can unquestionably absorb an attack and then retaliate are more stable than ones that can retaliate only if they are used before the attack arrives. Mistakes, accidental initiation of nuclear war, and more rapid Soviet escalation, are all made more likely by any plans to launch on tactical warning. As the MM/MX system becomes more vulnerable, and particularly if this system (and hence the entire bomber/ICBM portion of the triad) should become vulnerable to Soviet SLBMs, these pressures to launch on tactical warning will increase.

The MM/MX force is unstable under all of our arms control options except finite deterrence. In the case of finite deterrence, however, both sides would be restricted to perhaps 500 ICBM warheads and 1000 SLBM warheads. If US and Soviet ICBM warheads were then deployed on single-warhead missiles, as is likely, the attack price by ICBMs on one another would necessarily be greater than one, due to imperfect reliability and accuracy. In such a strongly warhead-limited world, this high attack price should discourage any attack. Thus our present single-warhead MM II missiles, even in their present vulnerable silos, would be stabilizing. Fears of a Soviet attack in a crisis, and pressures to launch on warning, would be reduced.

Verifiability and C^3I

Silo-based ICBMs are probably the world's most verifiable strategic weapons. They are easy to observe from satellites, and thus numerical limits on them are verifiable by "national technical means," i.e. without assistance from the other side. Warhead limits on silo-based ICBMs pose more of a problem, because external visual observation cannot tell how many warheads a missile is carrying. However, MIRVed missiles must be tested before deployment, and these tests are observable by national technical means. Thus, the SALT II treaty (signed but not ratified) and the proposed START agreement assume for counting purposes that all missiles contain the maximum number of warheads that they have ever been tested with. For example, under SALT II all Soviet SS-18 missiles are presumed to be 10-MIRVed even though some of them might be older single-warhead missiles.

Because they are fixed at known locations, silo-based ICBMs pose fewer problems in assuring command, control, and communications than do mobile missiles or submarines. The MM/MX control system should be adequate to carry out a pre-planned retaliatory strike. Like the missiles themselves, however, the control system cannot be counted on to survive for long under wartime conditions.

Summary

Except under restrictive 'finite deterrence' arms control limits, US Minutemen and especially silo-based MXs will be vulnerable and lethal and thus destabilizing in the mid-1990s. If Soviet SLBMs acquire greater accuracy, they might simultaneously threaten silo-based ICBMs and bombers, leaving submarines as the only independently-survivable leg of the triad. US silo-based ICBMs are destabilizing not only because of their vulnerability but also because MM III and MX carry multiple warheads, and because MM IIIA and especially MX pose a lethal threat to Soviet silo-based missiles. Furthermore, their vulnerability implies that there would be pressures, in a crisis, to launch these missiles on warning. The current mode does have the advantages of high verifiability and reasonably secure command and control.

References

1. Harold Feiveson and John Duffield, "Stopping the sea-based counterforce threat," International Security, Summer 1984, pp. 186-202.

2. *Report of the President's Commission on Strategic Forces*, Chaired by Brent Scowcroft, Library of Congress, Washington, D.C., April 1983.

Part IV Chapter 2

Evaluation: Launch on warning or launch under attack

An obvious way in which to reduce the perceived vulnerability of fixed land-based strategic ICBMs is to launch them before the attacking warheads either destroy them or prevent them from being launched. This possibility has never been renounced by official sources and therefore remains an option which we should explore.

Except for the no-upgrade option just examined this option is probably the cheapest. It is likely to be quite an effective answer to vulnerability questions but it generates other serious concerns.

Range of Options

Consideration of early launch requires an examination of the various types of early launches that can be contemplated. Each type will have its own different drawbacks. There is no unique method of either launch on warning or under attack.

One of the extreme postures is to launch missiles only after nuclear detonations on one's soil confirm unambiguously that nuclear attack has begun. The other extreme would be to launch when clear warning of an impending attack was received. Such an early launching could even include what is usually called preemptive attack. The threshold for launch on warning could be set so low that it amounts to preemption or so high that it approaches launch on attack.

The first warning of an attack is likely to be our early warning satellites which detect Soviet launches via infrared sensors. These satellites are in geosynchronous orbit and they look for launches from the known land-based Soviet ICBM sites and for submarine launches from either the Pacific or Atlantic Oceans. Within minutes following that warning, there would be signals from our land-based radar systems picking up the Soviet missile stages as they rose over the horizon. The large early warning radar that comprise the Ballistic Missile Early Warning System are located north of the US mainland in Alaska, Greenland and Britain and would pick up the launches from the land-based Soviet ICBMs. The newer large PAVEPAWS phased array radars are located on our coasts and look for submarine launches. In the future it might be possible to have more warning sensors in space. This may be one of the spinoffs of the SDI research and development effort.

In addition to the actual threshold that triggers a response, there are also options regarding the exact form of the response. If the US launched missiles at the earliest sign of attack, it might launch only a few missiles so as not to be disarmed either by an inadvertent attack or by probes by the opposition. Options exist to launch the missile in either armed or disarmed state, with the ability to change the state while the missile is in flight. However the US currently plans to launch its missiles in an armed state, without the ability to disarm. Clearly the level of the response is related to whether or not the armed state can be changed in flight, the threshold of warning and the targeting of the warheads.

Utility of System

If ICBMs are launched on short time scales, after warning or under attack, what can they contribute to our security? Countervalue targeting to enhance deterrence is always possible since those targets remain in place during an attack. However the high precision of the fixed land-based ICBMs is not needed for this type of targeting. The existing, lower accuracy submarine systems are adequate for this mission. In addition since the requirement for deterrence is that the retaliatory strike be inevitable but not necessarily prompt, the submarine force might be adequate as long as it was survivable. The actual number of warheads needed for deterrence is not that large and the value of additional warheads follows a well known law of diminishing returns. The question then is what targets would be assigned to prompt high precision systems.

To use the land-based ICBMs for counterforce hardened targets would exploit their high precision. However the Soviet hardened silos would be difficult to target since the missiles they once contained would presumably be the ones used in the attack on our silo-based missiles. To attack the remaining Soviet missiles that had not been launched would require the ability to rapidly assess and retarget. We would have to communicate to our systems which Soviet silos still had missiles in them. If the Soviets had a sequence of firings we could not be sure that the silos would still be occupied after we decided to attack them. In addition during the course of the attack (if not in a precursor attack), the sensors that supply us with those data might be disabled by Soviet countermeasures.

Counterforce targets such as hardened command and control centers would be well matched to the capabilities of our ICBMs. We ignore here the classical arguments about the wisdom of attacking C3 centers. That is, do we want to disrupt the Soviet attack by attacking the centers or should they be left intact to aid in war termination? We merely point out that fixed ICBMs launched with minimal warning have the capability of destroying these centers and for some scenarios prompt strikes against C3 might be a task in our single integrated operational plan.

Analysis

All versions of launch on warning or launch under attack raise serious questions of command and control both in normal times and during times of crisis. These two situations might generate different types of problems. As the decision times get shorter the effective command loop must move lower down the chain of command. We are discussing here the actual players in the time urgent decision making process, not the legal question of authority. If it is necessary to let the decisions move lower down the command chain, it is likely to be done by preassigned legal designation of authority under specified situations. Some such devolution of authority is believed to exist already to prevent decapitation strikes at the head of the command structure. The changes necessary for a launch on warning policy would be an extension of that idea but to an even lower level of authority. It is likely to be implemented during any possible crisis that could lead to a temptation for the Soviet to try a first strike. Assuming that launch on warning becomes our policy, it is prudent to assume that the policy will be known to the Soviets. If our launch-on-warning policy is to be an effective deterrent, the Soviets must know that it is one of our options. That knowledge pushes us to delegate authority early in a crisis to protect against a Soviet move since there could be a window of opportunity during a crisis for a Soviet advantage if we did not delegate authority early enough.

Clearly there is a tradeoff between guaranteeing the survivability of the land-based ICBMs and the danger of unauthorized, inadvertent or mistaken launch. The longer the warning time the higher the level in the command loop at which the launch decision can be made, but it is unclear if a difference of a few minutes is significant in preventing an unauthorized or inadvertent launch. However, it could make a difference in preventing a mistaken launch since more redundant sensor systems come into play and a single or a small number of failures would less likely to initiate a response. The problem of responding to a false alarms from a warning sensor is minimized if one can afford to wait for the later signals from other sensors. As the warning time decreases that redundancy weakens and the probability of responding to false alarms increases.

Attempts to protect against these problems have given rise to the idea of launching missiles with the ability of changing the armed state of the warhead (3). That means one can launch the missile in either the armed or disarmed state and then change that status in flight. Essentially, this extends the decision time by perhaps tens of minutes, a small but not insignificant time. For example, if a mistaken warning (no attack is actually taking place) results in a launch, within a short time the mistake would become clear. No Soviet warheads would approach our soil. Our launch could then be aborted. We can envision more complex command structures that would protect against unauthorized or inadvertent launch by having command centers separate from the launch control centers and requiring agreement on arming instructions or allowing one center to override the other. Various types of majority logic with possible preference toward the disarm orders could be designed. Some who argue for this toggling of the armed state of a warhead consider this feature more important than improving the survivability of the land-based ICBMs. They claim that all warheads including those on survivable submarine strategic systems should have this installed.

The installation of a toggle on the state of the warhead raises an interesting question as to whether it is more desirable to launch initially armed or unarmed. The different options give different fail-safe results and give different incentives for the Soviet targeting planners. In particular, Soviet targeting of US C3 centers might be different for the two cases. Of course one could mix the initial states of the missiles to complicate the other side's targeting plans. Both options could be implemented with the final decision made during a crisis.

A major concern of any system that allows toggling the arming state of a missile is the code security necessary to prevent the opposition from learning how to disarm the missiles. Cryptographic systems to preserve the integrity of the missiles only have to have tactical or short time security. If the system is not already broken there are only 30 minutes to try for the necessary pass codes. Modern coding techniques are probably adequate for this task. However, we are reluctant to move in this direction because of the uncertainty about whether our codes could become compromised. We are further concerned that the other side might think they can defeat our cryptosystems. An error of overconfidence on their part would be extremely destabilizing. Note that for the submarine force the navy has been resistant even to the conventional interlocks, the PALs, that are on other strategic systems.

There, of course, remains the basic question of how the Soviets could defeat our warning systems. These systems would be subjected to intense countermeasures, if not outright attack. Many have argued that an attack on the warning systems is itself a warning. While that is true it is also too simplistic for the case in which the response to a warning is a nuclear attack. If the Soviets launched an all out attack and included the warning systems as targets, as they most likely would, then using the initial attack on the sensors as the threshold condition for launch on warning would work. The Soviets have more subtle options which can cause complex problems for a launch on warning policy. They have the option of a small nonnuclear attack aimed only at our satellite warning system. That could even be extended over a period of time and look more like attrition than a single wave of attack. We have only a few warning satellites in orbit and our ability to rapidly replenish them is in doubt. We would be in a quandary. Do we launch a nuclear strike after a few satellites are destroyed or do we wait as see what happens next, but with reduced warning time? There must be a level at which we would refrain from a nuclear response if the Soviet move was clearly limited and we did not have an unambiguous consensus on what the Soviets planned next. Such a policy of launch on warning is susceptible to this type of tactic which would reduce the policy's effectiveness by slowly shortening the available warning time. This would not affect a policy of launch under attack which does not need the early and most exposed sensors.

In summary, this option of launching on warning seems to raise numerous problems, some of which might have technical fixes but are likely to be politically unacceptable. In an extreme case where only the land-based ICBM forces existed one would have to seriously consider this option, despite the dangers it generates. With the survivable submarine force and other nuclear forces we do not need to feel forced to accept the risks of the policy.

Notes and reference

1. The installation of land and sea based cruise missiles is changing some aspects of the triad. However, the basic three legs remain.

2. The ability of a given weapons system to destroy a hardened target varies as the yield divided by the accuracy cubed. Hence a factor of two increase in accuracy permits an eightfold reduction in yield for the same kill probability. See Part V Chapter 3.

3. Garwin, Richard L., "Launch Under Attack to Redress Minuteman Vulnerability?" International Security, 1982, pp. 117-139.

Part IV Chapter 3

Evaluation: Terminal defense of silos

Rather than replacing our current ICBMs with more survivable systems one can choose the option of defending them. The US does not now deploy any systems to defend its missiles from attack, but is allowed to deploy one ground-based system under the Anti-Ballistic Missile Treaty. From a strategic viewpoint, defensive systems fall into two clear categories, area defenses and point defenses. Area defenses have the ability to protect all targets within substantial areas of a country. Such defenses must be based at least partly in space to intercept missiles early in their flight while they can still potentially reach a wide range of targets. For example, layered defenses based largely in space might protect the eastern seaboard of the United States from Soviet missiles. Point defenses protect single targets which have often been "hardened," that is reinforced against the blast, thermal radiation and other effects of a nuclear attack. Typical point defenses might protect the NORAD command center under Cheyenne Mountain in Colorado but would offer no protection to Denver. Point defenses do not require a very large range because they must protect only one specific target. They are frequently based on the ground near the target they protect or have some limited mobility. Point defenses might, for example, depend on sensors based on aircraft which take off on warning of an attack.

Because this study concerns itself with the survivability of the land-based leg of the triad and any defense with area capabilities opens other strategic issues, we will deal only with hard point terminal defenses and not discuss any space-based defensive systems. A second reason for avoiding discussion of systems based in space is that their architecture is still undefined and any discussion of their merits in protecting ICBMs would be speculative.

Point defenses protect only specific targets such as missiles based in hardened silos. They increase the enemy's uncertainty that a first nuclear attack could disarm the defended side and discourage him from launching a nuclear first strike that would be met by a major retaliation. True point defenses do not offer any protection against a limited retaliation following a first nuclear attack. Thus point defenses of land-based missiles are strategically equivalent to mobile basing modes or superhardened missile silos. They arguably stabilize the nuclear balance in a crisis situation, although they may still fuel an offensive-defensive arms race. Some extend this argument to area defenses by claiming that they serve only as insurance against a first nuclear strike and are not inherently destabilizing.

Point terminal defenses come in two basic classes. The first, *conventional defense*, consists of rocket-powered interceptors based near the target to be defended and launched to intercept missiles entering the terminal phase of their flight toward the target. Conventional terminal defenses may protect both hardened targets and softer targets nearby such as radar antennas near silo fields. They are thus designed to intercept incoming warheads at such an altitude that a nuclear detonation would do minimal harm to targets on the ground.

The interceptor may be guided toward its target by infrared, optical or radar sensors carried on board, or it may be guided by sensors based on the ground or on a mobile platform such as an aircraft. Incoming RVs could be destroyed by nuclear detonations. Destruction by conventional explosions or the launch of devices such as pellet swarms in the path of the targeted RV would require significantly better guidance. Because the time involved in terminal phase is on the order of a minute or less and RVs must be destroyed at least ten kilometers above the surface in order to protect soft targets the interceptors require high acceleration boosters to get them to their targets in time. Hardened silos require smaller "keepout" ranges of less than one kilometer so the demands on the acceleration of the interceptor boosters are less stringent. Other critical technologies for conventional interceptors include target acquisition and tracking systems such as defense site radars and airborne or space-based radar, infrared and optical sensors as well as systems for terminal guidance of the attacking interceptor. The defender must consider countermeasures that could destroy or confuse his defenses.

Terminal defenses deployed by the Soviets consist of high acceleration interceptors, at least some of which carry nuclear warheads. The US system once deployed at Grand Forks is similar. Interceptors are guided to their targets by radars stationed near their silos. The targeting radars are alerted to the presence of incoming missiles first by satellite warning of an enemy launch and then by early warning radars which include large phased array radars stationed around the perimeter of both superpowers and over-the-horizon radars which guard other lines of attack. The large phased array radars have the ability to track incoming missiles and may be able to pass targeting information to tracking radars located near defensive silos. In the past, such terminal defenses have been vulnerable to attack on the radars which guide the missiles to their targets. Large phased array radars are very soft targets and are vulnerable to nuclear attack. They are expensive, take years to construct and thus cannot be easily proliferated. The early US Safeguard/Sentinal system also suffered from a lack of computer capability for battle management.

Technical improvements which may make terminal defenses more effective and less vulnerable include interceptors which have on-board tracking capability, more sophisticated computers and software development techniques, mobile radars and infrared or optical sensors such as those to be tested aboard the Airborne Optical Adjunct experiment under the auspices of SDI. Since the ABM Treaty explicitly forbids development of mobile components of ABM systems, the legality of such tests depends on the detailed interpretation of what exactly constitutes a "component" of an ABM system and whether or not such systems are based on "new" physical principles. The Soviets have deployed a massive defense against aircraft and their newest surface-to-air air defense missiles may have a limited capability against incoming RVs from ballistic missiles. Any terminal defense is subject to direct attack, enemy countermeasures designed to fool it and to saturation by an enemy who simply proliferates his offensive warheads.

The second class of terminal defenses, *novel defenses*, involves defenses that are not based on rocket-launched interceptors. These defenses are discussed in the literature but have not yet been deployed. They include concepts such as dust defenses, fratricide of missiles induced by close spacing of silos, or powerful lasers based at the site of the silos to be defended. The general argument in favor of such systems is their low cost relative to traditional terminal defenses and their immunity to saturation by a single massive strike.

In many cases, novel defenses may produce significant damage to the population or environment, which argues against deploying them. For example nuclear detonations near the surface in a defense that works by fratricide produce large quantities of lethal radioactive fallout which might cause a significant number of civilian casualties.

Most novel defenses are designed to operate with very small keepout volumes so they defend only hardened targets and offer no protection to nearby soft targets such as command and control radars. In this sense they are less likely to trigger an offensive/defensive arms race because their use is clearly an act of desperation on the part of the defender. Some novel defenses also depend critically on warning of an attack to trigger their use and are vulnerable to a false alarm which might cause significant damage to the defender's own territory.

Another issue in designing hard target terminal defenses is that the defender must decide what to defend. If he decides to defend only a fixed percentage of his silos, he can concentrate defenses on those silos and significantly improve their effectiveness. Such limited terminal defenses obviously involve difficult strategic decisions about just how many land-based missiles and of what type are needed to retaliate against the Soviets following a nuclear attack. Ideally, limited terminal defenses would be deployed so that the Soviets don't know which silos are being defended and will have to target all silos as if they were heavily defended. This drives up the Soviets' offensive costs while reducing our defensive costs.

Partially effective terminal defenses of hardened silos are technically feasible at this writing. The Soviets have deployed terminal defenses around Moscow. The system employs nuclear-armed exoatmospheric Galosh interceptors and high acceleration nuclear-armed Gazelle endoatmospheric interceptors with large ground-based radars for target acquisition and tracking. It is vulnerable both to direct attack and saturation.

Strategically point defenses are comparable to other methods proposed to increase the survivability of the land-based leg of the triad such as mobile basing modes. Their effectiveness will depend on future treaty limitations on offensive weapons and on changes in the ABM Treaty to allow such options as deployment at more than one site and the use of mobile sensor platforms. It seems likely that terminal defenses will be expensive relative to alternatives such as rail garrison basing, but the exact costs cannot be estimated without postulating a detailed architecture for the system.

The ABM Treaty allows a single ABM installation to defend fixed ICBM silos. The deployment is limited to 100 interceptors, with additional restrictions on the number and types of allowed radars. In the 1974 Protocol that reduced the number of allowed ABM sites from two to one, the United States specified that its single site was located at Grand Forks, ND. Even though that installation was later deactivated, it remains formally treaty-accountable. (The large missile site radar is still operational and is being used as part of the early warning network.)

If the United States were to opt for a terminal defense, the logical place to locate it would be at at Warren Air Force Base where MX is deployed. This raises the question whether it would be permissible to dismantle the Grand Forks installation and erect a similar one at Warren. The 1974 Protocol grants each side a onetime right to exchange its single ABM installation for one of the opposite kind; that is, the United States is allowed to exchange its Grand Forks installation for a national capital defense while the Soviets can exchange their Moscow installation for one in defense of ICBM silos. However, the Protocol says nothing about exchanging one ICBM defense site for another of the same kind. When densepack was being proposed as a basing mode for MX, a terminal defense was considered by the Department of Defense as a possible backup option to improve the survivability of the system in case the Soviet offensive threat increased. This would have required a switch in the location of the ABM system. The Administration was prepared to argue in that instance that the switch was legal, and would presumably take a similar position if the issue were to arise again.

Deployment of mobile defenses based on new physical principles is forbidden according to the strict interpretation of the ABM Treaty, and any interpretation forbids conventional mobile defenses.

Part IV Chapter 4

Evaluation: Strategic diad

The preceding evaluations (Chapters 1-3) have dealt with the major flaw in the current deployment of ICBMs - their vulnerability to attack by enemy ICBMs now and possibly to SLBMs in the future. Chapter 2 evaluated an operational answer to this vulnerability and Chapter 3 assessed a possible defensive answer. Implicit in either of those options is the assumptions that land-based missiles contribute to our security and ought to be retained. In particular, land-based missiles have more secure command and control and greater accuracy than the other two legs of the triad at the present time. Both features are important if US forces must be able to threaten counterforce strikes, i.e., attacks against enemy military facilities in order to deter aggressive acts. Neither is vital, however, if deterrence requires only that the US be able to threaten unacceptable damage to civilian targets in retaliation for a nuclear strike. Some even feel that the land-based missiles contribute to instability because their high vulnerability and lethality may invite attack in times of crisis.

But are land-based missiles are required at all? Might the other two legs of the triad be sufficiently enhanced to assure deterrence even in the absence of the land-based legs? If so then the US might simply allow the aging Minuteman missiles to fade into obsolescence, or actively dismantle the land-based force, perhaps as part of an arms control agreement.

If the warheads now deployed on ICBMs were transferred to submarines or strategic bombers, they would have a higher probability of survival. That is primarily because the US routinely keeps about two-thirds of the submarines carrying submarine launched ballistic missiles (SLBMs) at sea and has about 1/3rd of its strategic bombers on alert. The non-alert bombers are especially vulnerable to an attack by SLBMs, which might arrive with as little as 10-15 minutes warning time. Even though the USSR could destroy a large number of warheads by targeting submarine and bomber bases with relatively few warheads, hence have a large leverage for the use of its warheads, it could still not disarm the US. Several thousand warheads would remain on the alert vessels.

As arsenals shrink, the main problem is that of having too many eggs in one basket. To enhance survivability under possible future reductions to finite deterrence levels, the few remaining submarines and aircraft might be spread among many bases. Submarines in particular might carry smaller individual loads. Furthermore, the US might keep a higher fraction of its bombers on alert.

A diad needn't consist only of today's bombers and SLBMs. A diad could be "enhanced", for example, by the addition of small ballistic missile submarines off of US coasts, air-launched ballistic missiles, or other new deployments. Indeed, the planned additions of cruise missiles to airplanes, surface ships, and submarines are enhancements of just this sort. They contribute new capabilities to the US deterrent force. For example, air-launched cruise missiles enable the bombers to fire further from their targets and help make those bombers less vulnerable to enemy air defenses. Numerous small submarines over US continental shelves may present a significantly different anti-submarine warfare problem from a few large submarines in the deep ocean, so that the ability to destroy one type of submarine would not necessarily imply the ability to destroy the other type.

Only the ICBMs are capable of destroying missile silos quickly and with secure command and control. Air-launched cruise missiles are highly accurate and can destroy silos, but require several hours to fly to their targets. Today's SLBMs are too inaccurate to destroy silos, although this will change with the 1989 deployment of the Trident II missile. Thus a US diad could be less lethal, hence less threatening to Soviet forces, than the present triad. Whether this would actually be the case depends on whether, and in what numbers, Trident II was deployed.

Land-based ICBMs are the most vulnerable and the most lethal leg of the triad. For both reasons, a bomber/submarine diad should be more stable than the present triad. However, this assessment depends on the specifics of how the air-based and sea-based legs are deployed. The most stable diad would be one that is the least vulnerable and the least lethal.

A diversified diad based not only on gravity bombs and large deep-water submarines but also on some combination of cruise missiles, small submarines, air-launched ballistic missiles, and other enhancements should be much more survivable than the present triad, hence more stable.

Launch-on-warning pressures to use weapons before they are destroyed would nearly vanish in a diad. Upon strategic alert (i.e. in a crisis), and certainly on tactical alert (i.e. after receiving word of Soviet launch), the bombers could take off and all submarines could go to sea, where they would be relatively invulnerable. The only pressure might be to launch SLBMs from submarines in port if they were attacked before they could get out to sea.

Verifiability depends upon the details of diad deployment. ICBM silos are relatively easy to verify so removing them does not necessarily simplify the task of counting the remaining warheads. However, enhancements in the form of cruise missiles are more problematic. The current START talks are proposing to assure that each bomber capable of carrying cruise missiles holds a prescribed number of them. Air-launched ballistic missiles might be counted in the same way. However cruise missiles designed for submarines or surface ships cannot be verified by such counting rules because large numbers of these missiles might be hidden on the vessels. Furthermore cruise missiles carrying conventional warheads cannot be distinguished from afar from those loaded with nuclear warheads. One solution to this ambiguity might be to ban nuclear cruise missiles from surface ships, and count all submarine launched cruise missiles as nuclear.

By dismantling its land-based missile the US might risk weakening its strategic posture to the extent that the Soviets might be deterred from aggressive action by the threat of highly controllable, accurate land-based missiles. As a price for this move, the US might ask reciprocal measures on the part of the USS.R.. The US might be willing to destroy its MX and Minuteman missiles, for example, only as part of an arms control accord in which the Soviets agreed to comparable reductions in its force of SS-18 missiles. The START proposals move in this direction but currently allow either side to determine its own mix of land-based and sea-based missiles provided the total ballistic-missile warheads do not exceed 4800.

An SLBM/bomber diad would present greater control problems than the triad, because of the greater importance of SLBMs in a diad. Communication with submarines is difficult and tenuous, and might be disrupted in a nuclear war. For this reason, submarine crews have some authority to launch nuclear weapons on their own, independently of outside command. Still many are reluctant to forego the assured control over ICBMs for the uncertain link with SLBMs. This enhances command and control survivability, but raises questions about a unauthorized launch.

With a diad the US would lose the secure command and control that is possible with ICBMs. Hence the diad is an acceptable option to those who feel the US security does not require it to threaten counterforce strikes. A diad might be more stable than a triad because the forces would be less lethal (unless Trident II is included) and more survivable. Verification problems will be increased if the diad is enhanced by sea-launched cruise missiles, although these small missiles complicate any scenarios for future arsenals in which they are included. The price to implement a diad varies greatly with the enhancements that might accompany it.

The only economic cost of implementing a diad would be the price of dismantling missiles, if no other changes were made in the triad. But if the warheads removed from the land were replaced by equivalent deployments on bombers and at sea, or if enhancements were added, the cost could be considerable.

Part IV Chapter 5

Evaluation: Land-mobile Midgetman

The land-mobile Midgetman has been dubbed an "arms controller's missile." When the Scowcroft Commission this missile in 1983, stability was its main justification: its single warhead would make it a less lucrative target and its mobility would make it a less vulnerable one than the silo-based, mulitip[le-warhead Minuteman and MX missiles.

Two mobile Midgetman basing modes are under study: *Random mobile*, in which hardened mobile launchers (HMLs) move randomly over 5 large tracts of military land in the southwestern US, and *dash mobile*, in which HMLs remain parked at Minuteman silos in the northern midwest, ready to dash on warning onto surrounding roads and farms. The missile itself is a lightweight single-warhead ICBM (Figure 1). Its characteristics are given above in Chapter 1, Table 1.

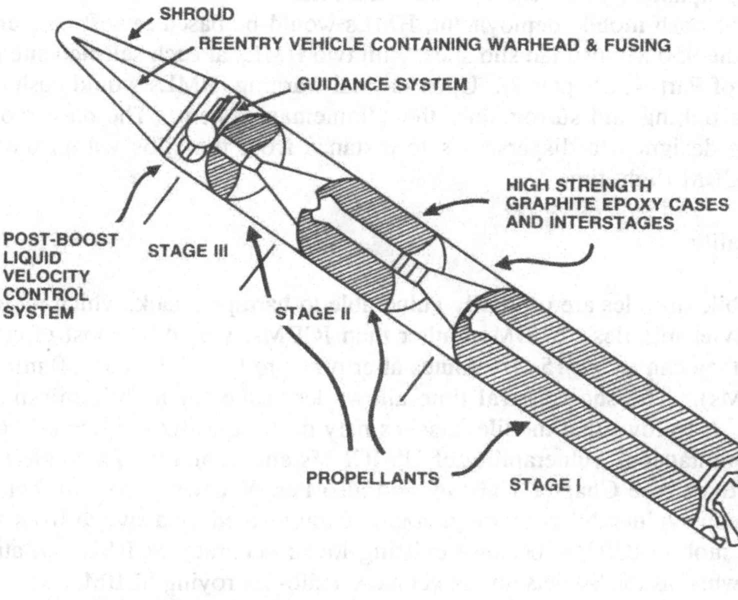

Figure 1. The Midgetman missile.

HMLs are special trucks designed to carry, protect, and launch one missile. They are designed to resist nuclear weapon effects by using armor and a low profile, and by lowering and anchoring themselves to the ground. An HML, according to scale-model tests, might withstand pressures that are 2 atm (30 psi) above normal atmospheric pressure. This strength reduces the dispersal area needed to protect the missile against barrage attacks.

The HML is pulled by a tractor and is designed for roads and for limited off-road travel. Its maximum speed on paved roads is 80 km/hr (50 mph). Its average speed was found to be 45 km/hr in tests of a full-size and full-weight HML over typical operating terrain and road. The total mass of the tractor, missile carrier, and 17 tonne missile, is 108 tonnes. Before launch, the carrier is placed on the ground in its streamlined configuration, and the tractor drives away; the missile is then fired remotely by a launch-control center. The carrier has retractable wheels and some method of anchoring to the ground to prevent overturning.

In the random-mobile mode, HMLs would be parked most of the time in the field, and randomly moved from time to time to deny an attacker knowledge of their locations. Deployment would be on 10,000 km^2 on the perimeters of several large tracts of Defense Department land in the southwestern US. During periods of increased tension (i.e. on "strategic warning"), the HMLs would expand their operating area to include the full 20,000 km^2 interior of these tracts. Upon receiving word that a Soviet attack is underway (i.e. upon "tactical warning"), the HMLs would dash off of DOD land onto adjacent roads and farmland, expanding the operating area still further.

In the dash-mobile deployment, HMLs would be based in soft bunkers at some of the 950 Minuteman silo sites, with two HMLs at each selected site (see Figure 7 of Part V, Chapter 7). Upon tactical warning, HMLs would dash onto the roads linking and surrounding the Minuteman system. The dash-mobile system is designed to disperse a safe distance from the silos within the 30-minute ICBM flight time.

Vulnerability

Mobile missiles are primarily vulnerable to barrage attack, which requires many Soviet missiles. SLBMs, rather than ICBMs, would be most effective because they can arrive 15-20 minutes after offshore launch (versus 30 minutes for ICBMs). The short arrival time allows less time for mobile missiles to disperse. It follows that mobile missiles may do little to fix the possible mid-1990s simultaneous vulnerability of US ICBMs and bombers to a single large SLBM attack (see Chapter 1 above, and also Part V Chapter 3). In fact, the simultaneous vulnerability problem could be aggravated by a switch from silo-based to mobile ICBMs, because existing lower accuracy SLBMs can attack mobiles whereas the Soviets do not yet have a silo-destroying SLBM.

Figure 2 shows the number of SS-18 ICBM *missiles* (not warheads) needed to barrage the areas over which 500 HMLs might disperse, as a function of dash

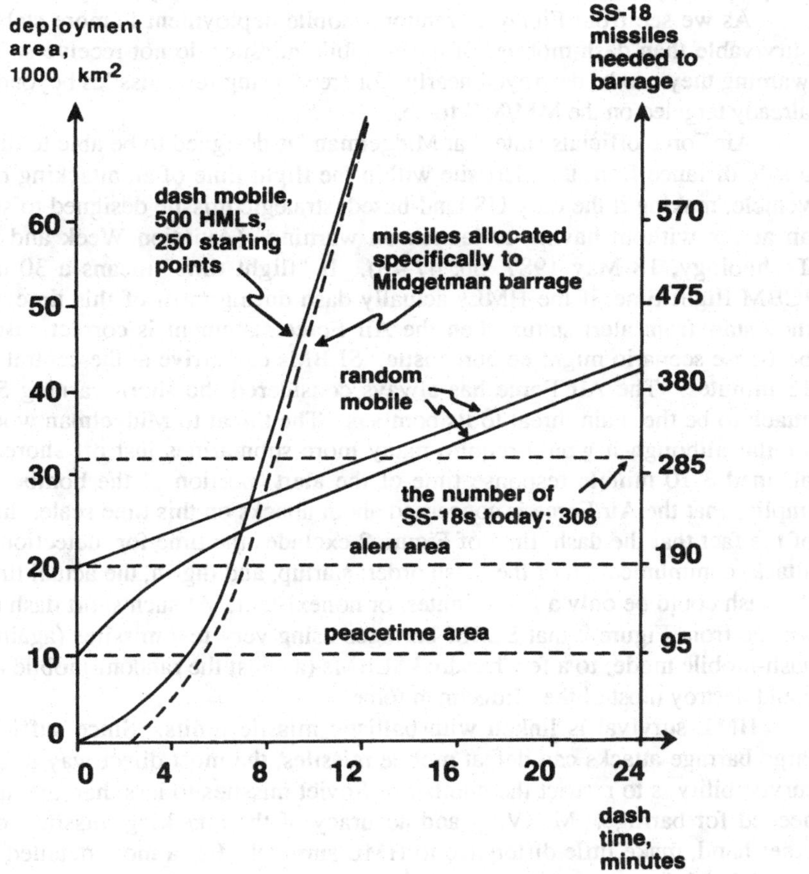

Figure 2. Deployment area generated, and number of SS-18-equivalents needed to barrage that area at an 85% HML destruction rate, as a function of dash time, for the dash-mobile and random-mobile basing modes. The attacking warheads are assumed to be air-burst at optimum height. The assumed dash speed is 45 km/hr (28 mph). "Dash time" means the time actually spent moving at this average speed. The total time to deploy, including warning, communication, start-up, and deployment in hardened configuration, would be longer. For dash-mobile, the dashed line gives the number of "excess" missiles needed, above those needed to target the Minuteman silos at which the HMLs are based.

time. The area exposed to 2 atm overpressure by a modern Soviet SLBM is about one third of the area exposed to 2 atm by a single SS-18 missile. Thus the graph may be scaled from SS-18s to SLBMs by multiplying by three.

As we see from Figure 2, random-mobile deployment is more stable and survivable than dash-mobile. If dash-mobile missiles do not receive sufficient warning they can be destroyed nearly "for free," using few missiles beyond those already targeted on the MM/MX force.

Air Force officials state that Midgetman "is designed to be able to disperse a safe distance from the alert site within the flight time of an attacking reentry vehicle, making it the only US land-based, strategic missile designed to survive an attack without having to launch on warning" (Aviation Week and Space Technology, 18 May 1987, pp. 47-48). If "flight time" means a 30-minute ICBM flight time, if the HMLs actually dash during most of this time, and if they start from alert status, then the Air Force statement is correct. But this best-case scenario might be unrealistic. SLBMs can arrive at the central US in 15 minutes. The Air Force has always considered the short-warning SLBM attack to be the main threat to its bombers. The threat to Midgetman would be similar although it would require many more submarines just off shore. The planned 3-10 minute response time of the alert portion of the bomber force implies that the Air Force is concerned about attacks on this time scale. In view of the fact that the dash time of Figure 2 excludes the time for detection of an attack, communication of the dash order, startup, and dig-in, the actual time for the dash could be only a few minutes, or nonexistent. At such short dash times, we see from Figure 2 that SLBM barrages using very few missiles (against the dash-mobile mode) to a few hundred SLBMs (against the random-mobile mode) could destroy most of the Midgetman force.

HML survival is linked with ballistic missile limits. Since sufficiently large barrage attacks can defeat mobile missiles, the most direct way to ensure survivability is to restrict the number of Soviet missiles to less than the number needed for barrage. MIRVing and accuracy of the attacking missiles, on the other hand, make little difference to HML survival. For a more detailed view, we study Midgetman survivability under each of our four arms control scenarios (unconstrained, SALT II, 50% cuts, finite deterrence).

Figure 2 shows that a force of 500 Minuteman in either the random- or dash-mobile mode can be destroyed by 300 SLBMs if available dash time is under 5 minutes or if the random mobile missiles are caught on their peacetime deployment area. Under SALT II constraints and certainly in an unconstrained world, the Soviets would have the required numbers of missiles to attack both these mobile missiles and US bombers. However the attack would require moving about 20 Soviet submarines close to US shores. Such a tactic would probably be detected and produce a high US alert level. Thus, if the Soviets wanted to have the capability of pursuing this tactic, they would probably need to deploy large numbers of submarines near US coasts on a regular peacetime basis.

At 50% cuts, if the Soviets maintained their present SLBM/ICBM ratio, they could launch a massive SLBM attack on Midgetman and MM/MX and the bombers only if all Midgetmen were deployed in the dash-mobile mode and were caught with under 4 minutes dash time. In a finite deterrence regime of 500 ICBM warheads (500 US Midgetmen) and 500 SLBM warheads on each side, the combined Soviet ICBM and SLBM forces would be unable to barrage the entire peacetime deployment area for mobile missiles., i.e. US missiles would be invulnerable.

Thus mobile missile survivability is ensured by great reductions in numbers of warheads. Other arms control measures that would enhance survivability include banning SLBM flight tests along depressed trajectories, so that warning times will not decrease even further, and a ban on testing of maneuvering reentry vehicles, to preclude ICBMs that could track down and destroy individual mobile missiles.

Lethality and stability

Midgetman will be about as lethal, per warhead, as MX. The limiting factor in the missile's lethality would then be its reliability. Reliability is likely to be lower than for silo-based missiles because of the required mobility. Still, the addition of 500 accurate warheads to the present MX and Minuteman IIIA and planned Trident II warheads increases the threat to Soviet ICBMs.

A major rationale for Midgetman has always been that its single warhead offers a less tempting and so more stable target than MIRVed missiles such as Minuteman III (3 warheads) or MX (10 warheads). The "attack price" (number of warheads used by the attacker, per warhead destroyed) quantifies this effect. Table 1 shows the prices to attack Minuteman, MX and mobile Midgetman.

Table 1. Price to attack (warheads used/warheads destroyed) various US ICBMs. Assumptions: The attack is by half-megaton warheads and is 90% effective. For dash mobile HMLs, only the "excess" SS-18s, beyond those used to attack the Minuteman silos, are counted.

Minuteman II (single warhead)		2.2
Minuteman III (triple warhead)		0.7
MX (ten warheads)		0.2
HMLs, dash mobile:	4 minute dash	0.0
	8 minute dash	2.2
	16 minute dash	21.0
HMLs, random mobile:	on peacetime area	2.1
	on alert area	4.2
	4 minute dash from alert	5.2
	8 minute dash from alert	6.2
	16 minute dash from alert	7.7

Midgetman's single warhead feature allows it to be light enough to travel. If MIRVed missiles could be carried in the random-mobile mode in an equally strong launcher, they would be as stable as random-mobile Midgetman. It is the deployment area and the total number of deployed warheads, rather than the MIRVing or non-MIRVing, that determines the attack price (and stability) of random-mobile Midgetman. Dash-mobile Midgetman, on the other hand, becomes proportionally easier to barrage as the number of HMLs goes down.

In a strategic alert, the mobile missiles must disperse to larger deployment areas. Some fear this move might appear threatening to the Soviets, and that there may be pressure on both sides to attack before mobile forces disperse. But others argue that dispersal would reassure the Soviets by making our missiles less vulnerable, and that dispersal of mobile missiles is no worse than the surge deployment of submarines during times of crisis.

Verifiability, C^3I, cost

Midgetman's mobility complicates verification. To confidently verify numerical limits on mobile ICBMs, cooperative methods going beyond national technical means will be needed. Since the Soviets have agreed to a variety of cooperative measures in the INF Treaty, it may be possible to get agreement on such measures in the case of strategic missiles as well.

A complete ban on mobile missiles would be the easiest condition to verify because observation of a single mobile missile would signal violation of the treaty. Allowed deployment of limited numbers might require designated deployment areas, with provisions for monitoring the traffic in and out and for conducting periodic on-site inspections.

Strict compliance with SALT II would preclude deployment of Midgetman (or of any other new ICBM) because it would be a second new type of ICBM as defined by the Treaty; each side is limited to one new type and the United States has already used its quota to deploy MX. Since the SALT II Treaty the Soviets have deployed two missiles. The Reagan administration has characterized one of these, the SS-25, as a violation of the Treaty. In the view of some critics, the Administration's charge has been less than conclusively demonstrated; the SS-25 was one of the major SALT compliance issues. But there is general agreement that the SS-25, even if technically not a new type, is perilously close to being one. If the US were to return to compliance with the Treaty, it might insist on a US right to deploy a second new type in response to the SS-25.

Land-mobile basing complicates the control problem, because mobile missiles cannot be connected to their launch control centers or to higher authorities by permanent, hardened land lines.

The control problem depends critically on whether the missiles are to be used only in a reflexive deterrent fashion, or in an active war-fighting fashion. In the deterrent mode, the control system is pre-programmed to carry out any one of several possible nuclear attacks, and the attack order is chosen from one of these existing plans. This places minimal demands on the control structure.

But war-fighting requires capabilities such as nuclear retargeting during a war, requiring a much more sophisticated and survivable control system. The low data rates of parts of the Midgetman communication system, the questionable survivability of HML crews, and the problem of C^3I survival, make it unlikely that adequate control can be maintained during a war.

500 mobile Midgetmen will cost about $50 billion, much more than the cost of other strategic systems. Table 2 compares the life cycle (acquisition, deployment, and 12 years of operation) costs of selected alternatives.

Since Midgetman's purpose is deterrence, and deterrence depends on survivable warheads, the cost per surviving warhead may be a more meaningful economic measuring rod. For example, a mid-1990s attack on 50 MX in Minuteman silos would leave only 25-50 surviving MX warheads, so the cost becomes $400-800 million per surviving MX warhead. It is difficult to compare this with Midgetman, because the size and plausibility of the Soviet attack are critical factors here. As one way of estimating this figure, a 90%-effective Soviet barrage of Midgetman would leave 50 surviving Midgetmen, for a cost of about $1 billion per surviving Midgetman warhead.

To reduce costs per warhead, there have been suggestions to MIRV Midgetman with 2 or 3 warheads. The 500 missile force could then be reduced to 250 or 167, with savings in HMLs and perhaps in missile costs. For the random-mobile (but not dash-mobile) mode, MIRVing would not affect warhead survivability under barrage attack, as long as the 250 or 167 HMLs are still deployed over the same total area. The added warheads would add to Midgetman's weight, and thus could reduce its mobility, but this might not be particularly important in the case of random-mobile deployment because the large deployment area is primarily a consequence of Midgetman's ability to move occasionally during peacetime, rather than its ability to dash rapidly during a crisis. Thus MIRVing could make sense, as a cost-saving measure, for random-mobile (but not dash-mobile) Midgetman.

Table 2. Life cycle cost estimates, in 1986 dollars, for several selected ways of deploying 500 warheads. Source: General Accounting Office, *Report to Congress on ICBM Modernization*, Washington, D.C., September 1986.

Method of deployment	Life cycle cost	Cost per warhead
50 MX in Minuteman silos (current)	$21.0 billion	$40.2 million
50 additional MX in MM silos	6.6	13.2
50 MX, rail mobile basing	23.3	46.6
500 Midgetmen, random mobile basing	52.1	104.2
500 Midgetmen, dash mobile basing	44.8	89.6

Part IV Chapter 6

Evaluation: Rail-mobile MX

The MX program was intended to give the United States the ability to hold the hardest Soviet targets at risk and to provide a highly survivable land-based strategic ballistic missile to replace the silo-bound Minuteman III. Early plans called for the missile to be deployed in a "race track" system with multiple protective shelters (MX-MPS), the missile being moved periodically from shelter to shelter. With "luck" and very good mimicking of the missile's signatures, the US could prevent the Soviet forces from knowing which shelter was occupied. Thus the Soviets would have to shoot at least two warheads at each one. Twenty shelters were contemplated for each missile.

The MX-MPS system foundered for several reasons. It was opposed by environmentalists, arms controllers and those concerned about the size of the budget for strategic offensive arms. The MPS system occupied too much land, required the land to be used only by the Air Force, might have had serious effects on the water supply in generally arid regions, was extremely expensive per surviving warhead, and gave the appearance of being a first- strike weapon because of its lack of manifest survivability. An ICBM that is both able to strike the missile silos of a potential enemy and that is itself vulnerable is a "shoot it or lose it" weapon. Since the missile force cannot ride out an attack and survive for later discriminate use, its owner must plan on using it before it is attacked -- and its opponent must regard the missile as having been constructed for executing a first strike attack. A major study called MX Missile Basing, conducted by the Congressional Office of Technology Assessment in 1981, failed to identify any other basing schemes with the requisite manifest survivability which also incorporated the robust communications possible for land-based systems. That study did examine rail-mobile missiles, but with the unstated assumption that such missiles would be on the rails at all times. According to OTA, rail-mobile missiles were survivable at any foreseeable level of the Soviet arsenal, but in peacetime would be vulnerable to accidents, sabotage and terrorism.

Congress allowed the Air Force to deploy the first fifty MX missiles in silos as direct replacements for Minuteman III missiles, but insisted that no further missiles would be procured until a survivable basing mode was developed. The rail garrison MX program was initiated in response to that challenge. The rail garrison deployment system envisions 25 trains, each with a complement of two MX missiles and each train being permanently stationed on existing Strategic Air Command bases. Between 7 and 17 SAC bases would receive the missile trains.

Survivability

According to Air Force figures, three hours after a decision to sortie from garrison was made, the system could not be barraged by the current Soviet SS-18 inventory assuming a yield of about 500 kt (kilotons) per warhead. Four hours after sortie, the force would be safe against even the largest projected increases in the SS-18 force. These figures assume a 50-80 km/h speed for the trains and minimal startup time (possible since the trains will use diesel locomotives) from "track alert" status. "Track alert" would be the rail-garrison version of putting bombers on strip alert where the aircraft are manned and ready to take off but without engines running. It is assumed that the first bifurcation in the available track occurs where the spur from the garrison meets the main line, a distance of not over 15 km from base. From that point on, track is generated at least twice as fast as the train moves since there is at least a twofold ambiguity as to the course of the train.

The Air Force arrives at its safety figures for the rail-garrison system by assuming that the readily achievable hardness of a loaded rail car is 5 psi. Above-ground nuclear weapons tests conducted before the 1963 Limited Test Ban Treaty demonstrated conclusively that even wooden boxcars were usable after experiencing 6 psi overpressures. A single 500 kt warhead, exploded at optimum altitude, can produce at least a 5 psi overpressure along a 7-km length of track. Each SS-18 could possibly have a war loading of 12 reentry vehicles. This is two in excess of the SALT II limit. However, in some tests SS-18s executed 10 real and two additional simulated releases of RVs. Assuming that the Soviet Union has confined its deployments to the ten-warhead SALT II limit reduces the length of track which can be barraged by a single missile. All forms of a Strategic Arms Reduction Treaty (START Treaty) now under consideration would, in fact, cut the SS-18 force roughly in half while permitting the deployment of at least 100 MX missiles.

Stability

The safe dispersal of the entire rail-mobile system still requires several hours. When the survivability of a weapons system depends on warning time, the enemy may be encouraged to strike soon rather than wait. It might perceive the vulnerable missiles to be either allocated to first strike or operating in a launch-under-attack mode. The President can defuse the situation by ordering the trains to sortie from their garrison. The President might be reluctant to give this order for fear of raising tensions in a delicate situation. On the other hand, the move could be viewed as a reassuring step. Missiles that are unlocatable and "untargetable" can be considered part of a secure second-strike reserve. Furthermore, because of the vibration and dislocation involved in moving a missile on the rails the guidance system of a rail-mobile missile is likely to be less accurate for several hours after a major movement than would be the same system housed in a silo. This could add to a general perception that scrambling the trains was a purely defensive and stabilizing gesture.

Verifiability

Some argue that rail-mobile missiles complicate the verification of arms control agreements. These points are raised: rail-mobile missiles can roam all over the country, be hidden in tunnels and buildings or be concealed in trains that resemble civilian trains. But rail-mobile missiles are just that -- confined to the railroad net. Rail lines are easily recognized features when seen in satellite photography even at the 10-meter resolution provided by the French SPOT 1 satellite. The Soviet rail net is particularly easy to inspect, since it is a skeleton whose back bone is formed by the trans-Siberian and BAM rail lines stretching across 7 time zones. Trains on tracks are identifiable in high-resolution imagery, but such pictures necessarily cover only fairly narrow angles of view. Hence, they cannot be used effectively to search for trains which have been ordered to sortie from garrison. Furthermore, the missile train could not be distinguished from any other train from a satellite photo.

Trains cannot be hidden for long periods in tunnels: Concealing a missile there would block an entire rail line, and construction of special tunnels would be observed. Trains cannot circulate at random, but must follow schedules -- particularly on a single track main line such as the Trans-Siberian -- since trains in opposite directions can only meet where sidings exist. Rails rarely enter buildings; they stop at loading docks.

Finally the facilities at which missiles are mated to rail cars will be very distinctive and easily recognized in satellite photography. Relatively straightforward analysis based on imagery of such a facility can provide good estimates of its maximum throughput and hence an upper limit to the number of deployed rail-mobile missiles. The number of such facilities in each country can readily be made a part of any future arms limitation or reduction agreements.

Since the signing of the INF Treaty in December, it is clear that verification does not have to be conducted solely by national technical means (NTM), in effect from satellites. Cooperative means of inspection and verification will shortly become the rule in strategic arms control agreements. Under such a regime, portal-perimeter monitoring of the rail- missile integration facilities could provide a highly accurate count of the number of missiles deployed on the rails.

Public acceptance of rail-mobile missiles would not be forthcoming if such a deployment meant that nuclear-armed missiles would be intermingled with ordinary rail traffic in ordinary times. The garrison feature of rail garrison MX directly addresses that problem. Despite Pentagon models of American rail-mobile MX missiles concealed in ordinary box cars, moving in ordinary freight trains, such a deployment is unlikely. The National Command Authority is unlikely to give the order for the trains to sortie except under the highest states of alert, designated DefCon 1 or DefCon 2, when the United States anticipates that a nuclear attack is imminent. Such high states of alert are apt to occur several hours before a premeditated attack occurs. While the garrison facilities are usually shown as simple structures, they can be built as "horizontal silos". One source estimates a hardness of 2000 psi for such structures, giving the garrisoned trains a reasonable chance of survival. The garrison could be designed to permit missiles to be fired after an attack.

Rail garrison ICBMs, generically, provide a nearly indestructible deterrent force because they can "generate trackage" in which the location of the train becomes uncertain so rapidly that they cannot be barraged. The track itself, together with the buried fiber optics communications systems which parallel most of the U. S. rail network, could provide communications even under the conditions of a nuclear attack.

Costs

The personnel costs of rail garrison missiles are low compared to those for truck mobile systems. Railroad equipment is vastly cheaper than, for example, any proposed version of a hardened mobile launcher. On the used market diesel locomotives can be bought for a few hundred thousand dollars or less; new they would cost somewhere around one to two million dollars. No HML with a hardness of 35 to 100 psi is competitive with the costs of the rail equipment needed to transport a missile. For MIRVed rail-mobile systems, garrison basing is very nearly as cheap, per warhead, as present silos.

But the rail mobile system, by itself, does not address the problem of vulnerability in case of a "bolt out of the blue," or "Pearly Harbor" attack. Under those circumstances, many missiles in garrison will probably be destroyed. Although nuclear bolts out of the blue seem improbable, rail mobility might need to be complemented by other basing modes -- such as superhard silos or smaller missiles transported by trucks -- for a limited number of land-based missiles, if one seeks absolute assurance that a fraction of the land-based leg of the triad will survive.

Part IV Chapter 7

Evaluation: Midgetman or MX in superhard silos

Another way to increase the survivability of US ICBMs would be to deploy them in heavily reinforced silos placed in high strength rock formations. These silos would have appreciable steel volume fractions and would exhibit considerable ductility (i.e. ability to deform without failing) when subjected to nuclear blast. Silos using common engineering materials should have a maximum static overpressure survival limit of about 3000 atm (45,000 psi). Current silo strengths are typically 130 - 200 atm (2000-3000 psi).

Silos might be able to survive an even greater *dynamic* overpressure if the shock arrives very quickly. The blast wave from a nuclear airburst has a virtually instantaneous rise time and a fall time of several milliseconds. When subjected to such short-lived strains, silo materials exhibit greatly improved mechanical properties because there is so little time for dislocations and local failures to interact with one another and cause a general failure. Thus, the Defense Nuclear Agency estimates that silos might withstand 7000 atm (or about 100,000 psi) overpressures. Such silos are called *superhard silos*. They are constructed with inner and outer steel shells with the space between the shells filled with steel bracing and high strength concrete. Such silos have much larger volume fractions of steel than found in current silos and would be placed in much stronger geological formations to reduce damage by cratering. Silo design is discussed further in Part V Chapters 9 and 11.

The destruction radius of a ground burst 0.5 MT warhead against a superhard silo is about 75 m, much smaller than the 300 m destruction radius of the same warhead against a Minuteman silo. 75 m is also the radius of the inner crater of an 0.5 MT warhead detonated in very hard rock. Although silos cannot survive within the inner crater, it appears that they can survive right up to the edge.

Current Soviet ICBM CEPs (i.e. inaccuracies) of 300 m (1) are expected to improve to 90 m (see Part V Chapter 3) in the 1990s. SLBMs are less accurate. The Trident II SLBM"s CEP is reported to be 120 m (2). Soviet SLBMs should achieve these accuracies in the 1990s. As Hobson discusses in Part V Chapter 3, the probability of silo survival when attacked by a single warhead may be approximated by:

$$SSPS = \exp(-RD^2/2\sigma^2) = 0.5^x \quad \text{where } x = RD^2/CEP^2.$$

Using CEPs representative of current and future ICBM and SLBM technology and crater sizes that could be associated with these weapons, we calculate the single shot probability of survival of superhard silos presented in Table 1.

Table 1 shows that superhard silos are currently invulnerable even to the most lethal missile in the Soviet arsenal--the 10-warhead SS-18. They are likely to remain invulnerable to attacks by SLBMs due to limitations on SLBM accuracies and throwweights. The Soviets can threaten superhard silos only by deploying heavy ICBMs with greater accuracies (CEPs near 90 m) and carrying fewer warheads of higher yield (greater than 2 MT).

Table 1. Single-shot probability of survival of superhard silos under attack by present and future ICBMs and SLBMs.

			crater radius (m)			fraction surviving	
Missile	No. of warheads	yield (MT)	ground burst	sub-surf. burst[a]	CEP[b] (m)	ground burst	sub-surf. burst
SS-18 ICBM	1	20	200	380	300	.70	.27
	10	.5	66	120	300	.96	.90
Future heavy ICBM (MX accuracy)	1	20	200	380	90	.04	.00
	2	7.1	145	280	90	.17	.00
	4	2.5	110	200	90	.37	.03
	10	.5	66	120	90	.68	.27
Future SLBM (SS-18 accuracy)	1	3.7	120	230	300	.90	.67
	2	1.3	90	170	300	.94	.81
	6	.25	54	100	300	.98	.92
Future SLBM (Tr II accuracy)	1	3.7	120	230	120	.50	.08
	2	1.3	90	170	120	.69	.27
	6	.25	54	100	120	.87	.61

Notes:

a. The penetration depth of the warhead was taken to scale as $(\text{yield})^{1/3}$ with a penetration depth of 5 m at a yield of 0.5 MT. This is basically an assumption of geometrical similarity for warheads of varying yields, and seems reasonable.

b. The 12 m CEP for SLBMs corresponds to the Trident II missile, while the 90 m CEP on the heavy ICBM corresponds to the MX accuracy.

Table 1 makes the unrealistic assumption that the attacking missiles and warheads are 100% reliable. This can be approximated if two warheads are used against each target (with each warhead being from a different missile to guard against missile and bus failures). If the missiles and warheads have reliabilities of 80% the likelihood of both warheads failing is about 4%. The first warhead explosion will have full effectiveness, but the resulting radiation burst, atmospheric disturbances, and debris may damage or deflect the second warhead, reducing its effectiveness.

Table 1 shows that, if earth penetrating weapons are deployed, the vulnerability of silos increases substantially. Subsurface explosions can have far larger crater radii than ground burst explosions and blast waves with far longer mechanical time constants, subjecting the silo to far more intense bending and shaking stresses than a ground burst explosion. Penetrator warheads will probably have a lower yield to mass ratio than corresponding air burst weapons because of the requirements for high cross-sectional mass density and structural robustness. This would require a much larger uranium content of the weapon and a corresponding increase in the fission fraction of the yield, which would in turn slightly reduce the relative advantage of penetrator warheads. The effects of earth penetrator warheads on silos are further discussed in Part V Chapter 9.

Table 1 assumes that the attacking missile is guided inertially. Terminally guided ICBMs would make superhard silos quite vulnerable. Furthermore, with terminal guidance the large yields indicated in Table 1 are not required. Even very small-yield warheads can destroy superhard silos if the accuracy is just a few tens of meters. With terminal guidance, SLBMs would be effective superhard silo killers as well. Depending upon the mass of the maneuvering system, such low-yield systems could use more highly fractionated warheads and increase the vulnerability of land based systems to a first strike.

The combination of terminal guidance with earth penetrators provides the attacker with the ability to destroy silos or deeply buried command centers. Indeed, the use of timed, simultaneously detonated, earth penetrating warheads in a hexagonally close packed pattern with the crater radii just touching allows the launching of a rock crushing plane wave down onto a deep hard site. The plane wave would attenuate to just below the rock crushing threshold and would then propagate as a lightly damped elastic wave. If the radius of the pattern of craters is comparable to the depth of the hard site below the surface, there would be little attenuation of the blast wave by the time it reached the hard site. The blast wave would reflect off the free surface of the hard site, doubling the stress to well above the rock crushing threshold, and destroying the center. Building hardened sites to survive such attacks requires taking advantage of relatively infrequent geological formations and strata as well as the use of specially adapted construction geometry and technique. Current sites (such as NORAD) may be quite vulnerable to such an attack.

The crisis stability of this deployment mode depends upon the missile: The MX's 10 warheads make a very valuable target, whereas Midgetman's 1 warhead and is one-tenth as valuable. While the Soviets would probably view an

MX to be worth 2 high-yield, high-accuracy warheads, they might not view a Midgetman as being worth such an expenditure. The ratio of the number of warheads used to the number destroyed (the "price to attack") is greater than 1 against Midgetman, but definitely less than 1 against MX.

Superhard silos are eminently verifiable. No other construction remotely resembles a superhard silo. The construction of a field of superhard silos with their associated access roads and control centers cannot be mistaken for any other activity. The silos can be easily counted, and would be assumed to have a missile in them. However SALT forbids the construction of new fixed ICBM silo launchers, or any modernization of existing launchers that increases their interior volume by more than 32 %. Superhardening would be permitted under SALT II only if it complied with these restrictions.

Silo basing allows time-urgent control of the missiles. They would have high data rate communication channels with their launch control centers, allowing rapid retargeting. They could be launched whenever desired, with no time required for deployment of the launching equipment.

Both of these missiles are designed to be highly accurate and have hard target kill capability. As with all silo launched missiles, the launching position is known accurately. Therefore the trajectories to attack chosen targets would be known. Both missiles can carry 0.3-0.5 MT warheads with CEPs of about 90m. The MX's high throw weight allows it to carry fewer warheads of higher yield, if desired to increase its lethality against hard targets. It could for example carry a single 20 MT warhead.

There are no current public estimates for superhard silo costs. The MX's large size requires a correspondingly larger and more expensive superhard silo than does the Midgetman. The mass of the MX silo and components would be five to ten times the mass of corresponding Midgetman components. On the other hand, 10 times as many superhard silos would be needed for Midgetman deployment as for an equivalent (in warheads) MX deployment.

The MX has been in production for several years and has amortized its development costs. The cost of producing additional missiles is less than $100 million each. Midgetman is still under development and would require development and testing before deployment. Thus the Midgetman missile itself is substantially more expensive than the MX missile.

If large numbers of Midgetman missiles and silos are produced, the cost per unit would drop substantially. Midgetman silo components are small enough that they could probably be factory fabricated and trucked to the site, something that would be far more difficult with MX components. Factory assembly of major Midgetman silo components could allow major savings.

Either Midgetman or MX in superhard silos would have improved crisis stability and launch control as compared to the mobile modes or launch-on-warning postures. According to Table 1, superhard silos are not excessively vulnerable to SLBMs, with their short flight times. Thus superhard silos would preserve synergistic survivability of ICBMs and bombers (see Part V Chapter 3). They would be vulnerable only to heavy ICBMs, which have a flight time of 30 minutes. Midgetmen in superhard silos are individually much less valuable targets, hence less likely to be attacked, than MXs.

Under START-like arms constraints of some 6000 warheads per side with throwweight limitations, it would be possible, but expensive in warheads and missiles used, for the Soviets to destroy 50 superhard MXs. It would not however be possible to destroy 500 superhard Midgetmen.

A ban upon the development of earth penetrating warheads would substantially increase superhard silo survivability. This could be accomplished by banning nuclear weapons tests above 1 kiloton, thus preventing the development of the hardened primary charge needed for earth penetrators. Earth penetrators would require substantial testing to assure survivability upon impact in hard rock at some 4 km/sec.

Similarly, the development and testing of terminally guided ICBM warheads needs to be restricted. Fixed basing modes have no survivability if terminally guided ICBM warheads are tested and deployed. If such deployment is not forbidden by treaty, superhardened silos have no strategic value.

References

1. Barton Wright, *Soviet Missiles: Data From 100 Unclassified Sources*, Lexington Books, Lexington (1986).

2. Joel Wit, "American SLBM: counterforce options and strategic implications," Survival, July 1982, pp. 163-174.

Part IV Chapter 8

Evaluation: Bunkered mobile Midgetman

The Midgetman missile is small enough that one could consider making cheap semi-hardened bunkers in which the mobile missiles could hide. This is a modification of the shell game once considered for the MX. The difference is that there exists commercial technology that can mass produce bunkers of Midgetman's size. One has to consider the tradeoffs between the number of bunkers, the strength, and the occupancy factor.

This class of deployment modes is specific to Midgetman because of its small size and mass. Unlike the much more massive MX, Midgetman can be easily moved by conventional equipment. The only land based transportation system that can carry the MX is the rail network.

Mobile deployment modes rely upon the multiplication of aim points to exhaust the attacker's warheads. Land-mobile Midgetman carried in hardened mobile launchers (HML) of 2 atm strength (Chapter 5) is one example. As another example, if land-mobile Midgetman's HML carried explosive excavation charges, it could dig itself into the ground, reducing its exposure to the dynamic blast that can overturn the HML and allowing its survival at overpressures of 10 atm. The carriers would have radiation exposures of 10^5-10^6 rad, if small warheads were used for the barrage. It is likely that the radiation resistance of the electronics would be the limiting factor in the hardening of dug-in mobile launchers.

The land requirements for the dispersal area of either of these mobile modes depend upon the strength of the protective carrier. The minimum land required would be just greater than the total area that could be covered with high enough overpressure to destroy the mobile missiles. Actual land area would exceed this estimate, but this estimate provides a good basis for calculation. The approximate land area that can be barraged by 2000, half-megaton warheads using two overlapped hexagonal close packed barrage patterns is given in Table 1, as a function of the carrier strength. Optimal detonation heights are assumed for the 2 and 10 atm overpressures. Ground bursts are used for the 100 and 1000 atm overpressures. The Soviets have many more than 2000 half-megaton warheads but would be unlikely to use more than 2000 to attack 500 single-warhead US missiles.

Table 1 also shows the land requirements of two other mobile deployment modes that will be the focus of this chapter. They rely on a multiplicity of relatively hard protection sites rather than on the carrier to provide most of the protection. The first uses specialized carriers sheltered in pre-built, heavily

Table 1. Minimum land requirements for deployment of 500 small missile warheads, as a function of the strength of the missile's protective device, assuming an attack by 2000 half-megaton warheads.

System	strength (atm)	area barraged (km^2)
Land-mobile Midgetman in HML	2	14,000
Dug in Midgetman	10	2,200
Semi-silo	100	380
Cheap hard silo	1000	80

reinforced, lateral *semi-silos* of 100 atm strength. The use of concrete and earth shields the electronics and the crew against radiation. Upon tactical warning, carriers dash to or among the semi-silos. The second uses a large number of mass-produced low-cost hard silos of 1000 atm strength: *cheap silos*. For either of these two modes, the Soviets would have to target all the silos (or semi-silos) as it would not know before launch which silos held missiles. The silos would be hard enough and far apart enough that the Soviets would have to target at least one warhead on each.

Table 2 presents the vulnerability of semi-silos and cheap silos to current and future ICBMs and SLBMs. We see that cheap silos are survivable against even the high-accuracy MIRVed SLBMs (CEP≈120 m) that the Soviets could deploy in the future. The attacking missile would have to be highly fractionated to attack numerous deployment sites. Cheap silos also have a small but non-negligeable survivability against MX-type 10-warhead ICBMs with CEPs of 90m.

In both of these modes the attacker is faced with destroying a number of missiles distributed among a much larger number of hardened sites. Due to the hardness of the sites, barrage tactics would be useless. There would be little incentive to attempt a preemptive strike because the attacker would have to use several times more warheads than it could destroy.

Both systems are quite survivable if the number of attacking warheads is constrained by treaty. In the case of an unconstrained arms race it is of course possible to attack every site in the deployment fields. Significant accuracy improvement, perhaps based on terminal guidance, could allow the Soviets to deploy larger numbers of small yield weapons. The resulting increase in the numbers of attacking warheads would decrease the survivability of the system.

If the attacker was able to discriminate full and empty silos, these deployment modes would of course become vulnerable. The use of decoys to confound targeting sensors would make verification difficult, if not impossible.

One problem with these "shell game" deployment modes is the problem of verifying (for arms control) the number of missiles in the field without at the same time revealing the locations of the missiles. This can be accomplished by dividing the field into a number of partitions, say 10 to 20, and shuttling one set of missile movement equipment between the partitions. When the missiles in a given partition are to be verified all the silos in that partition would be opened for satellite inspection. The movement equipment would then cover the silos and transport the missiles between silos so that observing satellites would not be able to determine which silos received the missiles. The routine opening of the silos in one partition after another would allow the verification of the missile count in the field without ever exposing more than the missiles in one field partition to targeted attack.

Semi-silos might have two operational modes. In the primary mode the carriers would be deployed in groups of 2 to 6 in barns located in a field of semi-silos. This is the cheapest mode and provides the greatest peacetime security for the weapons. Upon tactical warning the carriers would dash to randomly chosen semi-silos for protection. The roofs of the barns might be periodically raised to facilitate counting of the missiles. Under international tension the carriers could pre-deploy into the field. They would have access to pre-laid communications lines so that field-deployed carriers could dash to a different semi-silo under tactical warning. The small spacing of the semi-silos (a consequence of their high strength) would minimize the distance traveled by the carrier.

Table 2. Vulnerability of two hard mobile options (semi-silos and cheap hard silos) to present and future ICBM and SLBM attacks.

Attack missile	no. of warheads	yield (MT)	CEP (m)	destruct. radius (m): against 100 atm semi-silos	destruct. radius (m): against 1000 atm silos	fraction surviving: assuming 100 atm semi-silos	fraction surviving: assuming 1000 atm silos
SS-18	1	20	300	1120	500	.00	.10
ICBM	10	.5	300	330	150	.36	.81
Future	1	20	90	1120	500	.00	.00
ICBM	2	7.1	90	790	360	.00	.00
(MX	4	2.5	90	560	250	.00	.01
accuracy)	10	.5	90	330	150	.00	.15
Future	1	3.7	300	640	290	.04	.52
SLBM	2	1.3	300	450	220	.21	.73
(SS-18	6	.25	300	260	120	.59	.90
accuracy)							
Future	1	3.7	120	640	290	.00	.02
SLBM	2	1.3	120	450	220	.00	.14
(Tr II	6	.25	120	260	120	.04	.52
accuracy)							

For arms control verification, carriers should have a distinctive appearance to be verifiable from orbit. Semi-silos might be designed so that a side-looking satellite with the proper orientation would be able to verify whether a carrier was present in a semi-silo. Given the wide variations in semi-silo orientation on the ground, it would take numerous observations of a silo field to determine the occupancy of the field.

Since both of these bunkered mobile modes rely upon prepared sites, the launching locations would be known precisely and high bandwidth communication channels would be available for time-responsive targeting. Thus the missiles would available for hard target destruction under warfighting scenarios.

The economics of these modes may be dominated by the cost of the silos and their transporters. If cheap silos can be deployed for $10 million each, and if five times as many silos as warheads are deployed, the cost of the deployment field would be of the same order of magnitude as the cost of the missiles, which would probably cost approximately $30 million in quantity production. At ten times as many silos as warheads, silo costs would dominate. The operational cost of the cheap silo system would be relatively modest, involving the maintenance of the silos and the occasional movement of the missiles within the partitions of the field.

The semi-silo system would use lower cost protective structures, probably on the order of $2 million each, on a much larger land area. It would require the use of as many special purpose carriers as missiles, with significant personnel requirements. The larger garrisons and personnel requirements for this mode would be likely to dominate the long term costs. If ten times as many semi-silos were built as Midgetman were deployed, the cost of the protective structures would be on the order of $10 billion to shelter 500 missiles.

Consider an attack on 500 Midgetmen dispersed among 2500 cheap hard silos. Although the silos are individually cheap, a full system would probably have a total cost of $20-40 billion. An effective attack upon the field would require attacking all the silos which might contain a missile. An attack upon such a field with high accuracy warheads would be highly dependent upon missile reliability, as in the case of an attack upon superhardened silos (Chapter 7). If the missile reliability were 80% it would be necessary to attack each silo with 2 warheads, requiring the expenditure of 5000 warheads. If 0.5 MT groundburst weapons with a CEP of 90 m were used, this attack would leave about 15% of the silos intact. Such an attack would require 500 heavy ICBMs with accuracies greater than those currently deployed. The use of heavier warheads would reduce the survival fraction but increase the number of missiles needed for the attack. A single wave attack with 2.5 MT groundburst weapons with a CEP of 90 m and a reliability of 95% would leave 5% of the silos intact, but would require 625 heavy ICBMs. If missile reliability were 80%, a double wave attack would leave 4% of the silos intact at a cost of 1250 heavy ICBMs.

These attack requirements should be considered in light of the current USSR strategic arsenal of 1400 ICBMs, of which only 308 are "heavy." Most of these missiles do not currently have the accuracy required for such an attack, so the attack would be less effective than these estimates and yet require most of the modern ICBMs in the Soviet arsenal.

Even with greater accuracies such an attack would probably not be attempted under current arsenals, let alone under the reduced arsenals that would result from agreement upon a START treaty. As in the case of the superhard silos, the use of earth penetrator warheads would increase the destruction range of the warheads and eliminate fratricide. This would increase the lethality of the attack and might allow an attack by moderately fractionated SLBMs.

An attack upon 500 Midgetmen dispersed among 5000 semi-silos has similar characteristics to attacks upon Midgetman dispersed among cheap hard silos. Accurate, highly fractionated SLBMs could carry out such an attack. As before, each site would have to be attacked, requiring a minimal expenditure of 5000 warheads. If the attacking missiles had a reliability of 95%, about 5% of the silos (and missiles) would survive. If the attacking missile reliability was 80%, 10,000 warheads would be needed to attack the field, leaving 4% surviving missiles. These attacks would require 500 and 1000 heavy ICBM equivalents (including SLBM contributions), respectively. As before, these attacks require many more warheads than those being destroyed. Once again, terminal guidance would allow smaller-yield warheads and hence an increase in warhead fractionation and a decrease in the number of attacking missiles.

If an appreciable fraction of the exposed carriers could be targeted from space prior to the attack, or discriminated by the attacking warhead, an appreciable portion of the field could be vulnerable to a much smaller attack. While the missile carriers are outside of their semi-silos, they are vulnerable to destruction at very low overpressures. As exposed transport equipment, overpressures of less than 0.3 atm would be sufficient to disable the carriers. An attack upon a semi-silo field could be expected to begin with a barrage using nearby SLBMs to destroy any carriers which were not seated in their semi-silos. If launched on depressed trajectories, SLBMs could have flight times as short as 5 to 15 min, with 10 to 15 min being more probable. This sets the time frame within which the carriers might have to deploy into or among the semi-silos. It would be necessary have unmanned operation of the system or to maintain rotating crews on alert status with the carrier engines warmed and ready to go at all times for at least a portion of the missiles in the field. The short time restricts the distance that the carriers could disperse upon tactical warning and requires that the carrier sheds and associated garrisons be dispersed into the field.

If terminally guided munitions are deployed, the weapons might have some rudimentary ability to distinguish if a semi-silo was empty or not. This would require the deployment of decoys and could make verification more difficult. There is little advantage to silo hardness under terminally guided attack. The only thing that silo hardness buys is reduced deployment area.

Part IV Chapter 9

Evaluation: Multiple silos

The preceding chapter studied multiple shelters for Midgetman, which could possibly be built for considerably less than silos of comparable hardness for the much larger Minuteman or MX. However one could also consider multiple silos for these larger missiles if those silos could be built into rock and derive sufficient strength from that rock. The shelters would be spread over about 1000 km^2, allowing about 600 m between silos. As with the systems discussed in the preceding chapter, the locations of the actual missiles would not be known to the Soviets so that they would have to target each silo separately. Thus the price to attack would depend only on the total number of shelters and the total number of warheads deployed; the degree of MIRVing would not matter.

We evaluate here a hypothetical system consisting of 500 warheads (on 50 MX missiles, 167 Minutemen or 500 Midgetman) that can shuttle randomly among 2500 shelters. This system's protection arises from the fact that, although the individual silos could be destroyed by a high-accuracy ICBM attack, all 2500 silos would have to be attacked in order to destroy a high percentage of the 500 US warheads. Assuming double targeting, the attack would consume 5000 Soviet warheads just to destroy the 500 US warheads, and is thus so large as to be implausible under most arms control scenarios.

The multiple silos option is quite similar to the old "multiple protective shelters" (MPS) system proposed for MX basing in the late 1970s by the Carter Administration. For a thorough discussion of this plan, see *MX Missile Basing* by the Office of Technology Assessment (Washington, DC, 1981). One variant of the MPS plan would have housed Minuteman IIIs in vertical silos (*MX Missile Basing*, pp. 101-103).

A critical difference between our multiple silos option and the old MPS plan is that the MPS plan called for 30 times more land area. At least 30,000 km^2 was needed for the MPS plan versus 1000 km^2 under the multiple silos plan. This raised severe security and public interface problems for the MPS system, problems that eventually led to the rejection of this system. These problems are greatly reduced under the multiple silos plan, especially if the entire 1000 km^2 area is on a military reservation.

Survivability

Survivability of this system depends on two critical assumptions. The first is that it will be feasible to construct 2500 silos having a strength of some 700 atm (10,000 psi), for example by drilling directly into rock, and at a feasible cost. Some of the hardness would also come from a canister in which the missile could be transported. The second is that deception will be successful, so that the Soviets will not know which of the 2500 silos are actually occupied by missiles. Both assumptions are at least plausible, although highly tentative at this point. Hard rock such as granite has compressive strengths higher than 700 atm, and bolting and grouting (filling wall cavities and cracks with cement or epoxy) should ensure survival at these blast overpressures. Deception should be feasible, using for example a system similar to that which had been planned for the MX/MPS deployment scheme.

Essentially all 2500 silos would have to be attacked in order to destroy a large fraction of the US warheads. The 600 m spacing between silos implies that a single half-megaton blast (roughly the yield of most Soviet ICBM warheads) could destroy at most one silo, since the radius of destruction of such a warhead against a 700 atm silo is only 170 m. Even the very large 20 MT warhead on the older single-warhead SS-18 has a radius of destruction of only 580 m against 700 atm silos; although such a warhead might destroy two silos, it would be a bad choice for the Soviets because the same SS-18 missile could have carried ten half-megaton warheads and destroyed some five silos instead of only two.

To destroy a very large fraction of US warheads, the Soviets would need to attack this system in two waves. This requires 5000 warheads. The predicted 2-shot probability of survival (see Part V Chapter 3) is 4% under pessimistic assumptions (90 m Soviet CEP, 0.5 MT yield, 90% reliability), 15% under intermediate assumptions (110 m, 0.5 MT, 80%), and 33% under optimistic assumptions (125 m, 0.5 MT, 70%). Thus, the Soviets could expect that this attack would destroy the bulk of the US warheads.

This 2-on-1 Soviet attack uses 5000 Soviet warheads and destroys only 425 US warheads (assuming a 15% survival rate), so the attack price is about 12 warheads used per warhead destroyed. Only in a completely uncontrolled warhead environment could the Soviets devote so many warheads to destroying so few US warheads. The attack price against multiple silos is even higher than against random mobile Midgetman, even if we make optimistic assumptions about alert status and dash time for Midgetman. For example, after a full 16 minutes of actual dash from the alert deployment region, the price to attack Midgetman is "only" about 8 warheads used per warhead destroyed (see Table 3).

Assuming SALT II constraints on numbers of missiles (but modifications of SALT II to allow these new silos), the Soviets would have to devote two-thirds of their ICBM warheads just to destroy the multiple silo system. The remainder of their ICBM warheads would be needed to attack the remainder of the US ICBM force (2000 more warheads, assuming a 2-on-1 attack against 1000 additional silos). The Soviets would trade their large ICBM force for most of our much smaller ICBM force, an implausible trade.

To save warheads, the Soviets might instead attack in only one wave of 2500 warheads. However, this destroys only 325 (65%) of the 500 US warheads, and still requires a high attack price of about 8 warheads used per warhead destroyed.

An important feature is that SLBM attacks against multiple silos are entirely implausible. 5000 high-accuracy (Trident II accuracy) SLBM warheads would be needed just to attack the multiple silos. This is more warheads than the Soviets currently have in their entire SLBM force (about 3200), and seems implausible even in an uncontrolled warhead environment. Even a single-wave attack with 2500 high-accuracy SLBM warheads seems implausible. Thus, multiple silos should be secure against a quick SLBM attack directed against both US ICBMs and bombers: "synergistic survivability" (see Part V Chapter 3) of these two forces would be preserved.

Which missile is most plausible for this deployment mode? As we have discussed, multiple silos gain their protection from the fact that an implausibly large attack is needed to destroy the bulk of the 500 US warheads. This is a consequence of the large number of silos, and does not depend on the number of missiles actually residing in those silos. That is, 167 three-warhead Minuteman III missiles, or 50 ten-warhead MX missiles, or 500 single-warhead Midgetman missiles (carrying a total of 500 warheads in each of the 3 cases) are all equally protected, so long as the total number of silos is the same (we are assuming 2500) in each case. This deployment mode exhibits neither the typical "warhead-exchange advantage" of the single-warhead Midgetman missile, nor the typical warhead-exchange disadvantage of the ten-warhead MX missile. This is one case in which MIRVing is not especially destabilizing, and in which single-warheads are not especially stabilizing.

However, there are practical reasons for preferring a low MIRVed missile such as Minuteman III over either MX or Midgetman. The problem with the MX is that it is likely to be too large and too heavy to be removed from a vertical silo, transported to a different silo, and installed. Problems of this sort necessitated plans for horizontal shelters under the old MX/MPS plan.

The problem with Midgetman is that a large number of missiles are needed and so a large number of transporters, and many moves between silos, would be needed to keep the 500 missiles in uncertain locations among the 2500 silos. The multiple silo mode seems best suited to a missile with an intermediate number of warheads, small enough so that the missile can be transported and installed in vertical silos easily, but large enough to carry at least two or three warheads and thus reduce the total number of transporter and transportation moves required.

Lethality and stability

MX warheads are by far the most lethal in either arsenal, although they will soon be rivaled by the Trident II warhead to be deployed beginning in 1989. MX warheads are far more lethal than present Minuteman III warheads. This will also be true of Midgetman warheads. The "single-shot probability of survival" (SSPS) of Soviet silos attacked by either MX or Midgetman warheads is only

2%, whereas it is 40% for an attack on Soviet silos by Minuteman III warheads. Single-wave attacks by MX or Midgetman have high destruction rates, whereas single-wave attacks by Minuteman III do not: 500 MX or Midgetman warheads would destroy 440 Soviet silos in a single-wave attack on 500 silos, whereas 500 Minuteman III warheads would destroy only 270 Soviet silos (assuming 90% reliability in both cases).

The combination of the multiple silo system with present Minuteman III missiles could be highly stabilizing. The high price to attack this system makes attacks against it essentially impossible in an arms controlled world, due to a lack of sufficient Soviet ICBM warheads. Only in an all-out arms race in which the Soviets had many thousands of additional ICBM warheads would such an attack be plausible. At the same time, the moderate lethality of Minuteman III warheads would not go far toward forcing the Soviets into a "use them or lose them" situation in a crisis.

The combination of multiple silos with MX or Midgetman missiles is less stable, even though US missiles would still be essentially invulnerable. 500 MX or Midgetman warheads could threaten 500 Soviet silos, housing for example most of their highly-valued SS-18 and SS-19 ICBM forces. Especially in combination with other high-lethality US warheads such as Trident II, these MX warheads could pose a significant incremental threat to the Soviet ICBM force. At the very least, such a threat makes it more likely that the Soviets would put their ICBMs on (or close to) launch-on-warning status at an early stage in a crisis.

The stability of the multiple silo mode is greatly enhanced by the fact that its survival does not depend on receiving advance notice of a possible attack. Unlike the mobile systems, there is no premium on surprise, no premium on quick Soviet pre-emption at an early stage of a crisis. There is no pressure in a crisis for US commanders to redeploy the system in a less vulnerable, alert configuration.

Verifiability, C^3I, costs

It is easy to reliably count silos, but it is not so easy to count a limited number of missiles moving between a much larger number of silos, especially if location uncertainty is to be preserved.

The monitoring operations that had been planned for the old Multiple Protective Shelters (MPS) plan could be used for multiple silos. The essence of the MPS monitoring plan was as follows: (1) Silos were grouped into "clusters" such that each cluster contained only one real missile. If 167 Minuteman III missiles are placed in 2500 silos, then each cluster would consist of 2500/167 = 15 silos. (2) A barrier was placed at each cluster in such a way that missiles could not move into or out of a cluster without being observed by satellite. (3) Periodically, all the silos in one randomly chosen cluster were to be opened for two days of satellite inspection, to ensure that only one real missile was in that cluster.

This example shows that, even without on-site inspection, multiple silos should be verifiable. More elaborate monitoring operations, involving perhaps on-site inspection and electronic labeling of missiles, might make it possible to verify this system even without grouping the silos into segregated clusters.

Another arms control issue for multiple silos is the fact that this basing would violate the (unratified) SALT II ban on construction of additional ICBM silos.

Control of this system should be no more difficult than for other silo-based systems. All 2500 silos would be connected to launch control centers or to higher authorities by permanent, hardened land lines. Security of this system is greatly facilitated by the fact that it is spread over an area of only 800 km2. This should make it possible to put the entire system on controlled-access military land, reducing any threats from terrorists, spies, and saboteurs. Putting it on military land will also reduce public-interface problems.

No detailed cost estimates have been made for this system, but it might be much less expensive than land-mobile Midgetman, perhaps comparable to rail-based MX. The silos might not be very expensive, perhaps on the order of $1 million per silo for MM III silos, and three times that (because they require three times as much volume) for MX silos. Minuteman silos, being smaller, would be cheaper. The 2500 silos themselves might thus cost a few billion dollars. This estimate is based on engineering experience with tunneling in hard rock, an operation that has some common features with drilling silos in rock. The cannisters will be a major expense. They are likely to cost on the order of $50 million each, totaling some $9 billion for 167 MM III cannisters. The cost of the roads, transporters, missiles (these will be free if existing MM IIIs or MXs are used), launch equipment, command and control, other infrastructure, and operations must be added to this. Of course, these costs could be large, but they should be far less than the corresponding costs for random-mobile Midgetman.

In summary, this system appears to have some promise for being survivable. It must be emphasized however that it will be survivable only if based in some 2500 silos strengthened to some 700 atm, and only if location uncertainty of the 500 missiles among the 2500 silos can actually be preserved. Detailed studies are essential in order to determine the feasibility of these two critical conditions.

Multiple silos appear to have the largest favorable warhead-exchange ratio of any basing mode considered in this study, a feature that is favorable for stable deterrence. It can also be one of the least lethal, if present MM III missiles are used rather than MX or Midgetman. Thus for both survivability and lethality reasons, Minuteman III in multiple silos may be more stable in a crisis than the other land-basing schemes. The system should not be difficult to verify, although its location uncertainty makes it less easily verified than ordinary silo basing. This system's violation of SALT II could pose problems. Control should be good, as good as other silo basing plans and better than the mobile basing plans.

Part IV Chapter 10

Evaluation: Deep underground basing

An alternative approach to the increased hardening of the silos is to bury the missiles so deep that even a blast directly on top of the missile location will not destroy the missile. In such a situation improvements in precision by the other sides missiles will not lead to any change in vulnerability. The difficulty is that the earth above the buried missile will be quite disturbed and the system has to have a mechanism to insure that it can get to the surface intact and in a timely manner.

Deep underground basing (DUB) for ICBMs is a concept that has been studied for over a decade without attracting much serious interest. Its objective is simple enough-- to achieve survivability by providing a thick layer of solid material between the missile and the surface. No matter how accurate an attacking weapon, its "miss distance" for a surface burst can obviously be no less than the depth of burial. A deeply buried object is not subject to air blast or to thermal or nuclear radiation, while other effects are attenuated by the intervening material. Thus a missile deep enough underground should be secure against destruction by an attacking weapon of arbitrary yield and accuracy. Depths up to 2000-3000 feet or more are envisaged in basing schemes that have been studied; at such depths, even earth penetrators should not be a serious threat.

The generic DUB system consists of an underground tunnel complex a few hundred miles in total extent, with partially constructed egress shafts reaching close to the surface. The shafts would be backfilled for increased survivability, and would be completed to the surface only when a missile was to be fired; thus in principle there would be no external indicators of their locations.

The most promising version of the proposal calls for locating the deployment under a mesa or similar geological formation, with the tunnel complex at the general level of the surrounding terrain. In this way both access and egress tunnels can be made horizontal or nearly so, greatly simplifying construction and logistical problems and reducing costs. (An alternative version that has received some study would locate the system in the permafrost of northern Alaska. This scheme has much in common with the conventional one but offers some unique features.)

Stored within the underground tunnel complex would be a self-contained power supply, quarters and life support systems for the required personnel, and all equipment and supplies needed to enable the system to endure after a nuclear attack, dig itself out, and launch its missiles on command. Personnel requirements could be reduced by automation, remote operation, and cross training.

Tunnel boring machines with the required characteristics exist, but they are very slow: an advance rate of 50 feet per day in soft dry rock is typical. Thus even with the egress tunnels partially pre-constructed the system would take a long time to respond to an order to launch. Studies have estimated that utilizing state-of-the-art technology, missile egress and launch might be accomplished within about 20 hours. A system of this kind would be designed to be totally self sufficient and to have an endurance time of at least a year. The missiles could be either M-X or Midgetman (or any other, including intercontinental cruise missiles.)

Although a DUB system could be assigned a portion of the SIOP, its slow response time limits its suitability for that mission. On the other hand, its capability for long endurance and great flexibility could make it attractive as a secure, survivable reserve force. From this perspective, DUB would be viewed not as a candidate for basing a major component of the US strategic force, but rather as an auxiliary force that would buttress US retaliatory capability and thereby strengthen deterrence.

For a response that is not time-urgent, high-data-rate communications are not required; the system could rely on through-the-earth low-frequency C3 systems, which are slower but more survivable than are hard-wire communications. More work is needed on the C3 problem.

An advanced deep basing concept recently proposed would provide a fast launch capability with the addition of multiple interconnected launch shafts pre-bored to the surface. After an attack, several missiles could be launched through the surviving shafts. It is claimed that the launch response time could thereby be reduced to as little as 15 minutes. This concept has however received only quite preliminary analysis.

The principal advantage of deep underground basing is survivability. The system would be insensitive to improvements in missile accuracy or even to the development of terminal guidance. Moreover, it would promote crisis stability: there would be little incentive to attack a DUB deployment in a crisis, nor would there be any pressure to launch the missiles preemptively. Its slow response time, which diminishes DUB's attractiveness as a first-strike weapon, is another stabilizing factor.

Several technical uncertainties are associated with the problem of post-attack egress and launch. Although the intent would be to keep the locations of the egress points secret, the system is supposed to be able to launch its missiles even if this entails tunneling through the rubble in the crater and near-crater region created by a nuclear detonation. The pre-dug shafts could be collapsed or displaced by ground shock. Although many experts in tunneling technology are confident that egress can be accomplished with existing equipment and techniques, the capability has not been demonstrated. The Limited Test Ban Treaty precludes a fully realistic test of the system in a post-attack environment, although underground tests could provide much useful information.

The system faces other potential problems. An enemy could try to attack a DUB deployment with soft-landing nuclear mines, fuzed to go off upon detection of the acoustic signature of an emerging missile. There is no obvious counter to such an attack, assuming that the mines can be successfully landed.

If the response of the system were to be controlled from a remote facility, its survivability could be compromised. The system would be no more survivable than the launch control facility.

One practical problem would be disposing of the debris produced in excavating the tunnels. The debris typically has from 1.5 to 3 times the volume of the excavated cavity. Pre-boring the egress shafts could simplify the problem, but provision would still have to be made for the debris produced during post-attack digout.

Some preliminary estimates conclude that DUB would be cost-competitive with alternative options, but these cost estimates are subject to high uncertainties.

The verification problems posed by a DUB deployment are similar to those associated with other mobile basing modes. The missiles could not be observed while deployed but would have to be counted before they entered the tunnel complex, with some provision such as a barrier to guard against illicit introduction of missiles above the permitted number. Verification measures based on sampling techniques are not applicable to this type of deployment.

If any US ICBM were deployed deep underground, a question could be raised as to whether such a deployment violates the ban on construction of new fixed ICBM launchers. Inasmuch as a missile based deep underground would have to be moved to the surface before being launched, it could be argued that the launchers are in fact mobile and are therefore not covered by the treaty restriction. A similar problem loomed a few years ago when the Reagan Administration proposed to deploy MX in a closely-spaced-basing mode (densepack.) The Department of Defense argued then that the densepack launchers should be considered mobile (and therefore Treaty-compatible) because the missiles would be encapsulated and capable of being moved, even though they would in fact remain fixed in normal practice. This position was disputed by some critics, but the issue became moot when the proposal for densepack basing was abandoned. A launcher based deep underground would appear to have a stronger claim to being considered mobile than does one deployed in a fixed densepack mode. Still, the legality of such a deployment might be challenged by the Soviets.

In sum, although DUB has some attractive features, many technical and logistical problems would have to be resolved before the system could be considered a serious contender for adoption. The Defense Nuclear Agency and several contractors have been conducting feasibility studies since 1976, but the program has not progressed beyond that stage and has recently been curtailed.

Part V

Research articles

Part V Chapter 1

Stability of nuclear forces

Barbara G. Levi and David Hafemeister

Introduction

In its 1983 review of the U. S. strategic modernization program, the Scowcroft Commission became the first high-level government study to consider the destabilizing aspects of land-based multiple-warhead missiles and to recommend specific steps to enhance stability (1). One of the three steps towards modernization which they proposed was to initiate "engineering design of a single-warhead small ICBM, to reduce target value and permit flexibility in basing for better long-term survivability." They pointed out that "a more stable structure of ICBM deployments would exist if both sides moved toward more survivable methods of basing." They further note that "...from the point of view of enhancing ... stability, ... there is considerable merit in moving toward an ICBM force structure in which potential targets are of comparatively low value -- missiles containing only one warhead."

The Scowcroft Commission proposed small ICBMs mounted on road-mobile vehicles that move from their garrisons. Such missiles, once dispersed from their garrisons, could only be destroyed by a pattern attack over the entire deployment area, which might require several times more attacking warheads than the number carried by the small ICBMs being attacked.

Although the concept of stability has many meanings, we confine ourselves here to the one implied by the Scowcroft Commission's concern -- crisis stability. Crisis stability would exist if, in a time of great tension between the superpowers, neither side felt it could gain by being the first to initiate a nuclear exchange. Pressure to "go first" might exist if one side felt its nuclear weapons were vulnerable to a first strike by the other, or if one side calculated that, with a first strike, it could destroy more warheads than it used. Thus both the invulnerability of a basing mode and the number of warheads carried by a given missile affect crisis stability.

A simple exchange model

We have developed a very simple model to estimate approximately what would be the military consequences of a first strike by one superpower on the other under various possible arsenal structures. Several more sophisticated models exist and have been used to similar ends. (See, for example, reference 2.) However, these models too often are black boxes. We intentionally chose here to present a simple model in order to make the calculations both transparent and accessible. If one disagrees with any of the assumptions we have made, he can replace them with his own. We can justify our approach because the uncertainties in these calculations are so large that the results can only serve in any case to suggest certain broad conclusions. We have compared our results to those in more sophisticated models and find them qualitatively similar.

In formulating this model we have narrowly restricted ourselves to considering only those weapons targeted on other weapons. For the purpose of calculating exchange ratios we ignore other facilities that very likely would be targeted. These include command and control centers, communication facilities, etc. Keep in mind that the attacking side would want to use some of its nuclear forces for these targets as well as hold a certain fraction in reserve.

There are no unique, definitive criteria with which to determine the crisis stability of a particular basing mode, or the size of the force structure. Four possible criteria for assessing stability are as follows:

•*Warhead exchange ratio* .

R = (warheads destroyed)/(warheads used)

This ratio measures the relative advantage of the attacker. A ratio much greater than one indicates that the side being attacked probably has multiple-warhead missiles in relatively vulnerable basing modes.

•*Net warhead gain (or loss)* .

F = (warheads destroyed) - (warheads used)

This quantity would indicate how the attack has altered the overall nuclear balance. It is related to R by

F = (R-1) x (warheads used).

•*Surviving warheads* . This parameter indicates the ability of either side to launch a second strike. If, for example, the attacker destroys 2 or 3 times more warheads than it uses, but its enemy still retains several hundred or more warheads, the attacker has not necessarily gained a meaningful advantage. The adversary can retaliate with a devastating blow (providing it can still control its forces).

•*Ratio of Surviving Warheads.* This ratio might indicate whether the attacking side has altered the strategic balance in its favor, if indeed any such comparison is meaningful.

We can understand these parameters better by adopting a simple model to calculate their sensitivity to various properties of nuclear arsenals. Our simple model will seek to calculate the four parameters defined above for a hypothetical first strike by one of the superpowers ("Red") on the other ("Blue") . We make the following assumptions:

•When Red attacks Blue, a fraction f_a of Blue's aircraft are on alert and f_s of the submarines are at sea and therefore safe from attack.

•Red has a total of W(red) warheads. They are carried to their targets by missiles with a reliability of r. Each warhead approaching the target has a certain probability of destroying a missile in a single shot, denoted by SSKP.

•Red would target each missile silo with two warheads to ensure high enough probability of destruction. Red would also spend two warheads on each port serving as home to nuclear submarines. It would probably aim two warheads at the runways of bomber bases and also detonate airbursts in a pattern of perhaps 14 warheads around the bases to try to destroy aircraft attempting to take off.

•Blue has L_l (blue) land-based missiles, carrying a total of W_l (blue) warheads.

•Blue has all its bombers distributed among a total of B_a (blue) bases. Bombers at base and on alert carry a total of W_a (blue) warheads.

•Blue has all its submarines or nuclear-capable naval vessels stationed at a total of B_s (blue) ports. Submarines at sea and in port carry a total of W_s (blue) warheads.

•If Blue deployed L_m (blue) land-mobile missiles, carrying W_m (blue) warheads, and if a significant fraction of those missiles were normally dispersed over a large area, Red would have to create a barrage attack over the entire deployment area to have a reasonable chance of destroying them all. A barrage attack would be similarly required for rail-mobile missiles. We assume that the number of warheads for Red to attack the mobile component of Blue's land-based forces is given by M. Let A be the area of destruction created by each attacking warhead, and AD be the area over which the mobile launchers are dispersed. Then the probability of destroying a mobile missile is roughly (r M A/AD). In the case of railroads, these areas are replaced with track lengths.

It should be noted that, in our notation, the capital letter L stands for launchers, W for warheads and B for bases. The subscript l stands for land-based, s for sea-based, a for air-based and m for mobile.

With the parameters as defined above, we can write an equation for the number of Red warheads required for an attack on all the nuclear forces of Blue. We get the following expression:

$$W_{used}(red) = 2L_l(blue) + 2B_s(blue) + 16B_a(blue) + ML_m(blue)$$

The warheads destroyed depend on the reliability of these weapons and the fraction of the warheads that are caught on base:

$$W_{destroyed}(blue) = [1 - (1 - r \times SSKP)^2] \times W_l(blue)$$

$$+ [1-(1-r)^2] \times [(1-f_s) \times W_s(blue) + (1-f_a) \times W_a(blue)]$$

$$+ r \times (M \times A)/AD \times W_m(blue)$$

Note that, for the attacks with two warheads on one target (missile silos, submarine bases and airbases), we have included a term to calculate the combined reliability of two warheads, each with an independent reliability r. For the attack on missile silos, we include the combined probability of "kill" with two warheads, each with a single-shot kill probability of SSKP (the single-shot probability of survival, related to SSKP by SSPS=1-SSKP, is defined and discussed in Chapter 3). For barrage attacks on mobile missiles on alert with no overlap of the destructive areas of each warhead, we approximate the probability of destruction by the total area of destruction caused by M x L_m(blue) warheads divided by the total area of dispersal.

Exchanges at current force levels

We can now put in numbers for some of these parameters. We take the US as our model of possible forces on the blue side, and the USSR as a model for the red side. For US forces, typically one third of the bombers are on alert at any given time, and about 60% of the nuclear-armed submarines are at sea. Thus $f_a = 0.33$ and $f_s = 0.60$.

These fractions of survivable warheads are believed to be lower for the forces of the Soviet Union because their bombers are not on stand-by alert, and their submarines are only at sea about 20% of the time. The preponderance of Soviet nuclear force strength lies in their land-based missiles, which will become more vulnerable as the US deploys more of its highly accurate MX and Trident D-5 warheads. Perhaps to correct this imbalance, the Soviets are adding to their forces the mobile SS-25 missile, which rides on transporter-erector-launchers and the mobile SS-X- 24, which will be carried on railroad cars.

Typical reliabilities for modern intercontinental ballistic missiles are said to be about 80 or 90%. Consistent with assumptions favoring the side undertaking the first strike, we take r = 0.9.

To simulate present-day missile accuracies, we assume a specific CEP (the radius of the circle around the target within which half of the missiles aimed at the target are expected to land, a measure of the inaccuracy) for the Soviet SS-18 land-based missiles. Each of the 10 warheads on these missiles is said to have a CEP on the order of 0.25 km and a yield of about 0.5 MT. With these parameters, assuming a perfect reliability, one can use the methods presented in Chapter 3 to estimate that the "single-shot kill probability" SSKP for destroying with a single shot a silo whose hardness is 2000 psi (typical of US silos) is about 0.59. When combined with a missile reliability of 0.9, the overall probability of destroying a missile with a single shot is then 0.53. For future accuracies we take the CEP of the MX missile, which is about 0.1 km, near the limit of what may be possible without terminal guidance. With these accuracies, and a yield of about 0.5 MT, the lethality against a missile silo would primarily be limited by the missile reliability.

Table 1. Soviet strategic forces (4,5).

	launchers	warheads per launcher	warheads	yield (MT)
ICBMs:				
SS-11	184	1	184	1
	210	3	630	.27
SS-13	60	1	60	.67
SS-17	139	4	556	.75
SS-18	308	10	3080	.55
SS-19	360	6	2160	.55
SS-24	5	10	50	.1
SS-25	126	1	126	.55
	1392		6846	
SLBMs:				
SS-N-6	272	2	544	.7
SS-N-8	292	1	292	1.25
SS-N-17	12	1	12	.75
SS-N-18	224	6	1344	.35
SS-N-20	80	7	560	.1
SS-N-23	48	10	480	.1
	928		3232	
BOMBERs:				
Bear A	30	4 bombs	120	1
Bear B/C	30	5 bombs	150	1
Bear G	40	6 bombs	240	1
Bear H	55	12 ALCM, bombs	660	.25
Blackjack	0	22 ALCM, bombs	0	
	155		1170	
TOTALs	2475 launchers		11,248 warheads	

The current strategic nuclear arsenals of the U. S. and Soviet Union are summarized in Tables 1 and 2. We use these as a guide for input to our simple model, with Red representing the Soviets and Blue the US We also consider the counterforce scenario described in reference 3.

Table 2. US strategic forces (4,5)

	launchers	warheads per launcher	warheads	yield (MT)
ICBMs:				
MM II	450	1	450	1.2
MM III	200	3	600	.17
MM IIIA	300	3	900	.335
MX	50	10	500	.3
Midgetman	0	1	0	
	1000		2450	
SLBMs:				
Poseidon	256	10	2560	.04
Trident I	384	8	3072	.1
Trident II	0	8	0	.475
	616		5632	
BOMBERs:				
B-52 G	98	12 ALCM	1176	
B-52 G	69	8 bombs	552	
B-52 H	96	20 ALCM, bombs	1920	
B-1	95	20 ALCM	1900	
	362		5548	
TOTALs	1978 launchers		13,630 warheads	

With the following values for the parameters, we find the results for an exchange with the current arsenals:

$L_l(red) =$	1392	$W_l(red) =$	6846
$L_a(red)$	155	$W_a(red) =$	1170
$L_s(red) =$	928	$W_s(red) =$	3232
		$W(red) =$	11,248

$L_l(blue) =$	1000	$W_l(blue) =$	2310
$L_a(blue) =$	34	$W_a(blue) =$	4956
$L_s(blue) =$	16	$W_s(blue) =$	5632
		$W(blue) =$	12,898

$$W_{used}(red) = 2 \times 1000 + 2 \times 16 + 16 \times 34 = 2576$$

$$W_{destroyed}(blue) = 0.784 \times 2310 + 0.99 \times .4 \times 5632 + 0.99 \times .67 \times 4956 = 7328$$

Thus the warhead exchange ratio for an attack by Red on Blue is about 2.8. The ratio of surviving warheads is 1.6. See Table 3. Whereas Red had 1650 fewer warheads than Blue before the attack, Red now would end up with over 3000 more than Blue. From those numbers the first strike looks advantageous to the attacker. But Blue would still retain over 5,000 warheads and could cause massive damage in retaliation.

Red gains most of its leverage by destroying a large number of bombers at base and submarines in port with a small number of its warheads. The leverage would be reduced if there were enough warning of the attack to allow more submarines to sail out to sea or more planes either to take off or to disperse to other air bases.

Although under the current arsenals, one side might appear to have a significant leverage for the use of its warheads, the attack still does not come close to disarming the opponent. That is because a significant portion of Blue's arsenal -- bombers on alert and submarines at sea -- is fairly survivable at the present time. Moreover, the first strike would by no means be what some term "surgical." It might kill tens of millions of people and would surely invite a massive retaliation. Strategic planners worry nonetheless that communications with these bombers and submarines might not be sufficiently secure to enable command and control over these forces to mount the retaliatory attack.

The numbers of warheads on each side, both before and after a first strike, are plotted in Figure 1. Scenario 1 represents the case of existing arsenals. The ratio of warheads used and the ratio of warheads surviving are plotted in Figure 2, where scenario 1 again represents the exchange with existing arsenals.

Table 3. Results of a first strike by Red on Blue

Type of arsenal		Blue side:		Red side:		Ratio of warheads:	
		before	after	before	after	used	surviving
1.	Existing	12898	5570	11248	8672	2.8	1.6
2.	Existing + 500 mobiles (M=2)*	13398	5920	11248	7672	2.1	1.3
3.	Existing + 500 mobiles (M=5)	13398	5694	11248	6172	1.5	1.1
4.	START treaty	6000	2566	6000	4624	2.5	1.8
5.	START treaty with 500 mobiles (M=2)	6000	2808	6000	3724	1.4	1.3
6.	START treaty with 500 mobiles (M=5)	6000	2582	6000	2224	0.91	0.86
7.	Minimum deterrence (2000 warheads)	2000	1123	2000	224	0.49	0.20

*M stands for the barrage ratio, that is, the number of attacking warheads allocated to each single-warhead mobile missile.

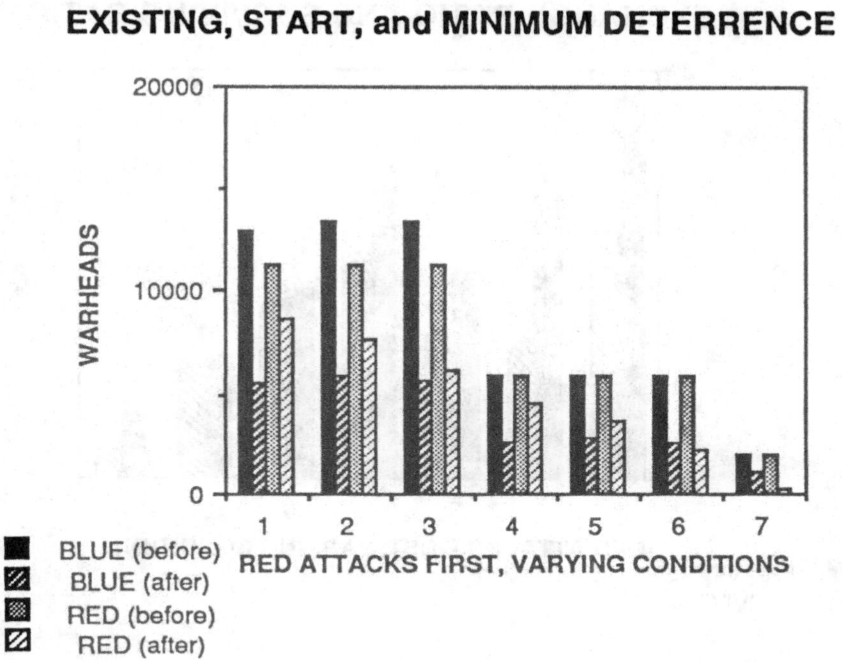

Figure 1. Histogram of the number of warheads in the arsenal of the Blue side before (black) and after (dark stripes) an attack by the Red side. The number of warheads in the Red arsenal before (grey) and after (light stripes) the attack are also shown. The depicted scenarios (also see Table 3) are: (1) existing arsenals, (2) existing arsenals plus 500 mobile missiles with Red barraging mobiles in a 2:1 ratio, (3) existing arsenals plus 500 mobiles with Red barraging mobiles in a 5:1 ratio, (4) arsenals reduced to 6000 warheads under a START Treaty with no mobiles, (5) arsenals reduced to 6000 warheads under a START Treaty with 500 mobiles and a barrage ratio of 2:1, (6) arsenals reduced to 6000 warheads under a START Treaty with 500 mobiles and a barrage ratio of 5:1, (7) minimum deterrence arsenals of 2000 warheads each with 500 mobiles and a barrage ratio of 2:1.

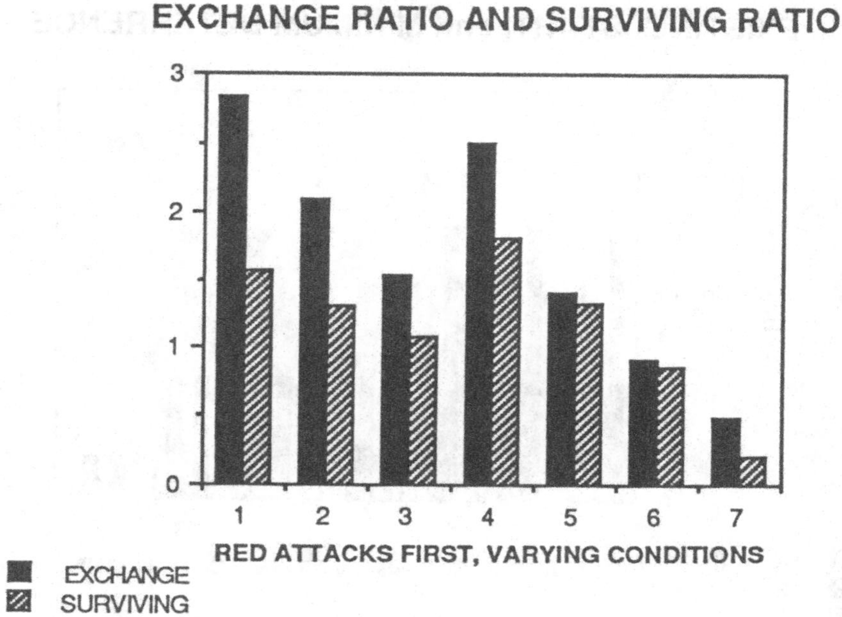

Figure 2. The histogram shows the ratio of warheads used to warheads destroyed (black) and the ratio of surviving warheads (grey) for an attack by Red on Blue. The scenarios are the same as those in Figure 1.

Exchanges at current force levels with the addition of mobile Midgetman

Let us now examine the impact of adding to this current force structure a more "survivable" component of the land-based missile force. In particular we look at the impact of a mobile land-based force that requires a number M of enemy warheads to destroy each launcher. One mode for deploying mobile missiles being considered in the US is to make them randomly move over a large peacetime deployment area of 10,000 km^2 (4000 mi^2). They would spread out to an area of 20,000 km^2 in times of tension. A Midgetman small ICBM is designed to withstand overpressures up to 2 atm (30 psi). It would be exposed to this level of overpressure or higher if it were within an area of =6.9 km^2 around an airburst warhead from a Soviet SS-18 (0.5 MT each). (The area would be 5.4 km^2 for a .35 MT warhead from a sea-launched missile.) Thus the Soviets would need nearly 3000 SS-18 warheads to barrage the entire alert deployment area, assuming no overlap of warheads. If the US deployed about 500 Midgetmen over this area, it would take about 6 warheads to destroy each

Midgetman. However, we feel that Red might not want to spend more than two of its own warheads on any one enemy warhead, so we studied the effect of varying the ratio M from 2 to 5.

The US currently appears to favor deployment of mobile missiles at existing Minuteman missile sites, with two mobiles stationed at a given Minuteman silo, ready to dash on warning of an attack. The survivability of mobiles in this basing mode would be highly dependent on warning time. If the mobiles do not have time to dash, they would be destroyed by the same warheads that target the Minuteman missile. If the mobiles can dash for 6 minutes at 50 km/hr (30 mi/hr), however, they would disperse over 20,000 km^2, an area equivalent to that of the random mobile mode.

In our model we assumed that the mobiles were deployed in a random mobile mode. The results apply to the dash-mobile mode with enough warning for a 6-minute dispersal. (See the treatment of this subject in Chapter 8.) If the time were so short that dash mobiles were caught at the Minuteman silos, the results of the attack would be the same as those for the case with no mobiles deployed, except that Red would destroy a bonus of 500 more missiles.

With 500 land-mobile missiles added to the current US arsenal, and Red targeted them with a barrage ratio M = 2, the results of a first strike are shown as scenario 2 in Figures 1 and 2. Scenario 3 shows the results if M = 5. See also Table 3.

The addition of 500 Midgetmen has decreased the warhead exchange ratio. While there remains an apparent advantage for Red to strike first, Blue would still clearly retain a sizable force. With a 2:1 barrage, Red would have an exchange ratio of 2.1. If Red undertook a 5:1 barrage, the exchange ratio would be 1.5.

Exchanges at proposed START levels

Suppose now that both sides cut their numbers of warheads roughly in half, to 6000 warheads, as proposed in the current START talks. We assume that the US might choose to keep 1900 warheads on 400 land- based missiles (100 MX and 300 of its most accurate Minuteman III missiles). It may wish to keep 2900 warheads on submarine-launched ballistic missiles -- nearly all its current deployment of 3072 Trident I warheads, to be replaced by Trident IIs. Then the US would have exactly the number of warheads allowed on ballistic missiles, according to some draft proposals for START. It would then have 1200 warheads on its bombers.

If the Soviets chose to stay within START limits but retain roughly the same ratio of land-to-sea based warheads, it might deploy about 3000 warheads on ICBMs and 1800 on SLBMs. It, too, would have 1200 warheads on its bombers. (Actually, under counting rules now under consideration, START would leave both the US and USSR. with many more than 1100 nuclear weapons deployed on its aircraft. That is because the short-range air missiles, air-launched cruise missiles and gravity bombs loaded onto strategic bombers would not count individually as one warhead each).

If the Red and Blue Sides in our model made similar decisions about their arsenals as the US and USSR are likely to under START, the exchange ratio for a Red attack of Blue would be 2.5. See Scenario 4 in Figures 1 and 2 and Table 3.

However, if START allowed mobile missiles and Blue deployed 500 Midgetmen rather than 50 additional land-based missiles, the exchange ratio for a Red attack on Blue using a 2:1 barrage of mobiles would be about 1.4, and using a 5:1 barrage it would be 0.91. See scenarios 5 and 6 in Figures 1 and 2.

Exchanges at finite deterrence levels

Finally we look at the situation with both superpower arsenals reduced to a total of 2000 warheads. If Blue deployed its 600 land- based warheads on mobile missiles, we calculate that Red, after attacking the submarine and bomber bases, would not have enough warheads to barrage the entire deployment area of the Midgetmen. If Red used all its arsenal, it might destroy by barrage only 216 of the Midgetmen along with 396 of the SLBMs at the 16 naval bases and 265 warheads at the 34 air bases. For these future scenarios we assume that the missiles would have greater accuracies than is generally true today. The results with a barrage ratio of 2:1 are summarized as scenario 7 in Figures 1 and 2. The Red side would retain only 224 warheads while the Blue would still have over half its original arsenal. Neither side would be able to perceive any advantage to striking first.

Note that, as arsenals shrink, the nature of the modernized component of the land-based forces has increasingly large influence over stability. In particular, the land-mobile Midgetman removes the advantage of striking first both because it requires more warheads to attack each launcher and each launcher carries only one warhead.

Even when a given scenario might produce a favorable warhead exchange ratio for the attack, a large number of warheads might still survive. If these warheads were targeted in retaliation on cities, it should take only a few weapons to deter, much less than the 400 equivalent megatons that is often cited as the criterion for a minimal deterrence. But if the warheads are to be used for "counterforce," or "nuclear war fighting," one might argue that larger numbers are needed. If one considers the effects of these weapons on urban populations, Blue would certainly retain more than the assured destruction level in all of the cases we have considered here. As the arsenals shrink to lower and lower levels, the number of warheads remaining after a hypothetical first strike decreases. But in all cases, Blue retains several times more weapons than required to destroy about 20% of Red's population -- provided the surviving weapons can be retargeted and launched.

Although our model is a very simple one, the exchange ratios are qualitatively similar to those from more sophisticated models (2). The advantage of a simple model is to be able to calculate the answers readily for any choice of parameters.

Rail versus road basing

Chapters 8 and 9 with the relative survivability of road mobile and rail mobile ICBMs. A calculation of survivability depends on warning or dash time, area or length available for dispersal, velocity of the missile carriers, hardness of the missile carriers, number, yield and pattern of attacking warheads. Even with this information, a calculation of mobile survivability is not as accurate as for the case of the silos. In the calculations so far we approximated the kill probability as the ratio of total area barraged to the total area of dispersal. In a more elaborate calculation, one might want to factor in the time factor -- how far the Midgetmen can dash in the flight time of the given missile.

As discussed by Peter Zimmerman in Chapter 9, Air Force figures indicate that the rail-mobile MX would be safe from attack by all the current SS-18 missiles if they had about 3 hours warning of an attack. One reason the trains become increasingly difficult to barrage is that the track branches and the trains might turn on any of these branches.

Conclusions

The results of our stability calculations depend very strongly on the overall size of the force structure. With more than 10,000 nuclear warheads in each of the arsenals of the US and the USSR, one side still has an advantage in striking first even if 500 new mobile missiles replace 50 MX missiles (10 warheads each) based in vulnerable silos, although the net gain is then only marginal. On the other hand, if the superpowers continue their current efforts to reduce the nuclear arsenals, and especially if they move toward mobile single-warhead missiles, they may increasingly move toward more stable arsenal configurations.

Of course, while these numerical measures of "stability" may measure one factor that could influence the action of a nation during a crisis, it is only one such factor. As long as the nuclear arsenals remain as large as they are, both superpowers must realize that any attack, even if restricted to the military targets of the other side, would kill tens of millions of civilians and surely invite an equally devastating retaliation on the attacker.

References and notes

1. *Report of the President's Commission on Strategic Forces*, chaired by Brent Scowcroft, April 1983, reprinted by the Library of Congress, Washington, D.C. (1983).

2. Michael May, George Bing, and John Steinbruner, "Arsenals after START: the implications of deep cuts," International Security, Summer 1988, p. 90.

3. William Daugherty, Barbara Levi, Frank von Hippel, "The consequences of 'limited' nuclear attacks on the US," International Security, Spring 1986, p. 3.

4. Robert Norris and William Arkin, Bulletin of the Atomic Scientists, Jan 1988, p. 56.

5. *The Military Balance 1987-88,* International Institute for Strategic Studies, London (1988).

Part V Chapter 2

Verification of limitations on land-based missiles

*Barbara G. Levi,**
David Hafemeister,
and Valerie Thomas

Introduction

Two factors have recently changed the nature of arms-control verification: The first is the trend towards small and mobile land-based missile systems, which are inherently more difficult to monitor. The second is the precedent for intrusive, on-site verification measures set by the Intermediate-Range Nuclear Force (INF) Treaty ratified in 1988. Indeed, to some extent the first factor begat the second: The treaty formulated on-site inspection measures to deal with the small and mobile missile systems addressed by that treaty. These unprecedented cooperative measures, however, only became feasible because of a new political climate between the superpowers.

When the SALT treaties were written, all land-based strategic US and Soviet missiles were deployed in heavily reinforced, underground concrete silos. These missile silos could be easily seen and counted by so-called "national technical means," that is, by monitoring entirely from outside the territory of the adversary. Negotiators for the SALT treaties did worry somewhat about the possibility that silos could be hidden or that missiles could be stockpiled for reloading in known silos, but the general consensus was that militarily significant violations of this sort are unlikely to go undetected by national technical means.

Those concerns are now more worrisome with the development of strategic missiles that are mobile, smaller and more easily concealed. These new ICBMs are being designed to elude attempts to target them, so they may also foil attempts to count them through satellite reconnaissance. Cooperative measures, such as on-site inspection of various facilities, may now be necessary to increase confidence in the verification by national technical means.

The INF Treaty has cleared the path for cooperative measures by giving a prominent role to on-site inspection (1). One might expect that the provisions for verification now being negotiated as part of the Strategic Arms Reduction Treaty (START) would include on-site inspection at least as stringent as those of the INF Treaty because the strategic weapons are regarded as more critical to national defense. The two verification regimes are not likely to be identical, however. First, the INF Treaty bans entire classes of mobile missiles whereas START might allow limited deployment of such systems. Next, the two sides may have different standards of verification in the strategic case. Finally, differences in physical characteristics, as well as details of production and deployment between intermediate-range and strategic weapons systems, may affect the feasibility of some verification approaches.

On-site inspection could give the public greater confidence in treaty compliance, independent of whether these measures provide significantly more information than that gathered by national technical means or by exchange of data. On-site inspection might also affect the overall relationship between the United States and the Soviet Union. The presence of inspectors from one nation on the soil of another could in itself improve the international political atmosphere. On the other hand, challenge inspections could have a counterproductive effect, by raising the level of tension and suspicion (2).

In this chapter, we will first outline various technologies and procedures for verifying compliance with possible arms control agreements on ICBMs. Next we will examine the characteristics of the possible new land-based missiles, and will consider what techniques might be useful in verifying compliance restraints. While our discussion focuses on verification of limits on the numbers of missiles being deployed, we will discuss limitations on other properties as well.

Types of verification measures

One approach to verification is the use of *national technical means* (NTM). These include any technical measures that allow one nation to observe the other from outside the territory of that nation. NTMs are completely under the control of the monitoring nation. Thus we have satellites flying overhead, or planes and ships cruising just outside national borders, all equipped with various devices for looking or listening in on the other nation. We also have access to information from seismographs recording nuclear testing signals. Some details of these devices are summarized in the Appendix A. Treaties such as SALT I and II and the (unratified) Threshold Test Ban Treaty rely principally on NTMs for verification.

With lots of attention now focused on verification, the capabilities of NTM may improve. After the INF Treaty was signed, the Senate Select Committee on Intelligence drafted a proposal for the US to deploy more highly sophisticated photo reconnaissance and radar-imaging satellites to keep closer tabs on mobile missiles (3). The proposal reflects the feeling of the Senators backing it that, while existing intelligence capabilities are sufficient to verify compliance with the INF Treaty, greater capabilities would be required to verify provisions of START.

Cooperative measures among the parties to the treaty can complement and enhance NTM. A mild form of such cooperation was the agreement by the Americans and Soviets, under the SALT treaties, not to interfere with the NTMs of verification by the other Party. Both sides consented not to use deliberate concealment measures such as camouflage, nor to encrypt any telemetry data from missile tests that was necessary to verify compliance with treaty provisions. To help the Soviets know which US B-52 bombers were carrying air-launched cruise missiles, for example, the American bombers had "functionally related observable differences." Such measures can be circumvented. However, the offending side would risk detection by NTM and the subsequent cost of getting caught.

On-site inspection is more intrusive, requiring foreign inspectors on sovereign soil. The Soviets have historically been more resistant than the US to on-site inspection, viewing it an opportunity for the US to spy. However, since the early 1980s, the Soviets have been increasingly amenable to some forms of on-site inspection. In 1985, the USSR agreed to allow the International Atomic Energy Agency to conduct inspections of some Soviet nuclear power plants, in conjunction with the Nuclear Non-Proliferation Treaty. Beginning in the summer of 1986, they allowed scientists sponsored by a private US organization, the Natural Resources Defense Council, to install and monitor seismographs at several locations around its nuclear weapons test site at Semipalatinsk (4). Most recently and most surprisingly, they agreed to the on-site inspection provisions of the INF Treaty.

Historically, the United States has favored on-site inspections. But now that on-site inspection has become a realistic option, concern for preservation of military secrets has moderated US enthusiasm.

Verification arrangements can have several complementary objectives. The US Department of State has defined these objectives as follows: to ensure confidence in the agreement; to deter violation of the agreement by increasing the likelihood that such violations would be detected; and to permit timely detection of violations, so that appropriate steps to protect US and allied security can be taken (5).

Provisions of the INF Treaty

The INF Treaty bans several classes of US and Soviet ground-launched nuclear missiles: intermediate-range (1000 to 5500 km) and short-range (500 to 1000 km) ground-launched ballistic missiles, as well as ground-launched cruise missiles. Its provisions seek to assure each party that the other side has destroyed all existing missiles and launchers, that it has effectively deactivated former missile operating bases and support centers, and that it is no longer manufacturing missiles of the forbidden classes. The treaty provides for short-notice inspections by each party to certain declared facilities on the other side. Provisions governing those inspections are summarized in Appendix B.

In addition to the short-notice inspections, each party to the INF will permanently station up to 30 inspectors around the clock outside one missile production facility of the other side to ensure that INF missiles are not being produced. The Soviets will monitor the Pershing II production facility in Magna, Utah. The US will monitor the final assembly facility for the SS-20 missile, where the SS-25 is assembled (6). The US was concerned that the USSR could use its continuing production of the strategic SS-25 missiles as a screen for illegal production of SS-20s, because the first stage of the SS-25 is externally similar to the first stage SS-20. The stages are reported to differ internally. All vehicles large enough to contain a proscribed missile, or its first stage, must exit only through one specified portal. At that portal, inspectors may weigh, measure and photograph the containers as they leave, or even look inside containers large enough to contain the first stage of an SS-20. Nondestructive inspection measures, such as x rays, may also be used.

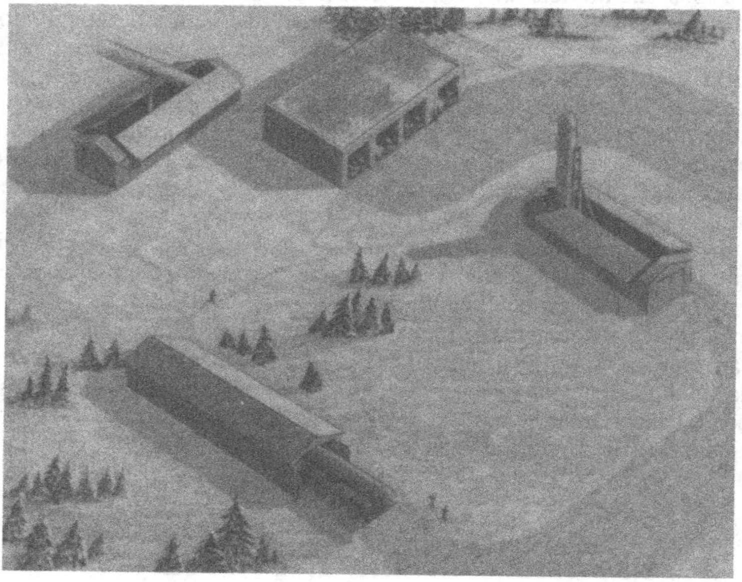

Figure 1. Artist's concept of the SS-20 or SS-25 missile with sliding-roof garages. [Department of Defense.]

Monitoring land-mobile missile deployment

Bases for the two Soviet land-mobile missiles -- the theater SS-20 missiles and the strategic SS-25 missiles -- have distinctive characteristics that have enabled easy identification of them by remote sensing. It is more precise to call them moveable rather than mobile, since they can be moved, but are not designed for continuous travel. The missiles are deployed at fixed, presurveyed sites. Each missile is stored in a garage-like building with a sliding roof (Figure 1). SS-25s are based in groups of three. According to one US diplomat, this is a basing scheme "we would not want to modify even if we had a right to do so" (7).

The locations of several Soviet missile bases were even identified by analysts from West Germany using the comparatively coarse 30-m resolution of the US civilian Landsat imagery (8). The key to their identification was the pattern of roads characteristic of SS-20 bases.

If approved, the US Midgetman small ICBM is expected to be similarly deployed in well defined areas -- either moving at random in its hardened mobile launcher over limited reservations (Figure 2) or garrisoned at the sites of existing Minuteman silos (Figure 3). Many support facilities are necessary to keep these missiles in an alert and functional status (9). A main operating base would provide the principal command, operational, security, maintenance, logistics and

Figure 2. Midgetman in its hardened mobile launcher. [US Air Force photograph.]

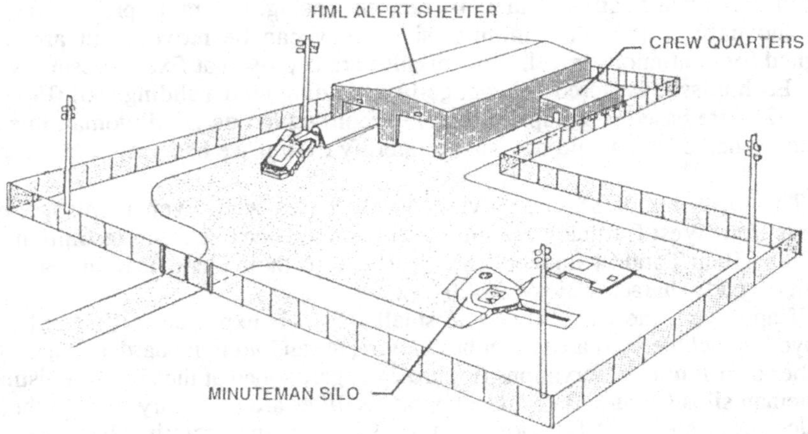

Figure 3. Drawing of two Midgetman alert shelters at Minuteman site. [US Air Force drawing.]

personnel support functions. Practical considerations require that a base be capable of accommodating at least 40 launchers. About 4,000 personnel is the estimated requirement for normal operation of a mobile force consisting of 200 missiles (10).

When the Midgetman traveled on public roads, it would follow required convoy procedures. Thus it is likely to be accompanied by two vehicles carrying security equipment and personnel as well as a communications van, and to have a helicopter hovering overhead. These missiles, although designed to be mobile, may not move with ease or speed (11).

A complete ban on mobile missiles would be easier to verify than a numerical ceiling on them. Each nation could use NTMs to search for mobile launchers and supporting bases. Even one missile sighted would be a violation of the treaty. Both the US Midgetman and the Soviet SS-25 are large enough and have distinctive enough characteristics to be identifiable by visible-light satellite imagery. For example, the launcher carrying the US Midgetman is expected to be 30 m long, 4 m wide, and 91 tonnes in weight.

Moreover, these missiles require considerable support systems, as described above, and might be detected even more easily by the bases from which they operate. Any clandestine attempt to deploy a militarily significant number of mobile missiles would be sufficiently extensive that it would be surprising if some part of it were not observed in time, by human intelligence, signal intelligence or satellite reconnaissance.

There is some concern in the US Congress that the Soviets may hide missiles produced before the treaty comes into effect. This issue was raised during the Senate Foreign Relations Committee's hearings on the INF Treaty. As part of the exchange of information required under the INF Treaty, the Soviet Union stated that it has deployed 405 SS-20s and has manufactured an additional 245 that are not deployed (12). These Soviet numbers conflict with estimates by the Defense Intelligence Agency, which has asserted that the Soviet Union has 410 to 545 nondeployed SS-20s. However, the Soviet numbers were not in conflict with the Central Intelligence Agency estimates (13). The two agencies use different methods to make their estimates. Some Senators fear that the Soviets may not be revealing all of their INF missiles. In response, Paul Nitze, the senior arms control adviser to Secretary of State George Shultz, said that the Soviets could maintain a clandestine, militarily significant force only at great cost and risk and with decreasing reliability. The ban on flight testing, which is readily verifiable by NTMs, would lead to increasing uncertainty in the reliability of any clandestine force (14). Nevertheless, this controversy indicates the difficulty of estimating missile numbers by NTMs alone. This topic will be addressed in the next section.

The prospects for successful verification of a complete ban appear good, especially when supplemented by on-site inspection. The bases for normal deployment of mobile missiles are large and distinctive. It would be difficult for either nation to deploy a significant number of mobile missiles in violation of a treaty banning such missiles and retain confidence in their performance.

Verifying compliance with a treaty limiting but not banning mobile missiles is considerably more complex than verifying compliance with a total ban. Each side would want assurance that only the allowed number of missiles had been deployed. Each side might want to monitor rates of production and assembly, and to keep track of the comings and goings of missiles and the vehicles that transport and launch them.

Restricting missiles to so-called *designated deployment areas* would simplify the task: one missile observed outside this area would constitute a treaty violation. Provision for designated deployment areas would not impose much inconvenience, because mobile missiles require too much infrastructure for effective deployment over an unlimited expanse. Treaty provisions might call for fences around production facilities, deployment areas, and maintenance and storage centers. Each nation might be required to give prior notice of missile movements between the facilities. Measures such as these are being discussed at the START negotiations (15).

A nation might circumvent the treaty either by building a clandestine base or by stockpiling missiles and launchers without any support structures. That nation would then either incur the high cost of increasingly complex command, control and security as well as a heightened probability of detection, or would have to accept a lower reliability for the additional missiles (16). Complex and intrusive monitoring raises the cost of cheating and increase the likelihood of getting caught, but clearly does not preclude it.

An alternative to construction of clandestine bases for prohibited missiles would be covertly to stockpile additional missiles near the base or missile maintenance facility and circulate them into and off of the bases for exercise and maintenance. Such movements might be attempted at night or under cloud cover. Infrared satellite imagery might be able to track the movements at night by the heat of the engines, and future radar satellites may have sufficient resolution to detect these missiles under cloud cover. Even stationary missiles and launchers may be at a different temperature than their background, and hence detectable at least in principle by infrared sensors. The extensive communications at the bases might provide additional clues to any clandestine activities.

Cooperative measures could make such violations much more difficult. The missiles might be equipped with identifications large enough to be "read" from space. All traffic might be required to move in and out of the designated deployment area by certain declared routes and through portals operated by inspectors. If the missiles were housed in garages, the two sides might consent to lift the roofs periodically to reveal the numbers of missiles inside. Under the INF Treaty, the Soviets are required, at US request, to open the roof of a specified SS-25 deployment base, and to remove all missiles on launchers from the buildings for display in the open. Up to six such requests may be made each year (17).

Challenge inspections might be used to increase confidence in treaty compliance. Neither side is likely to agree to "anytime, anywhere" inspections. A limited number of inspections at designated missile facilities and designated deployment areas may be more acceptable. The INF Treaty provides for challenge inspections of this type, which are described in Appendix B.

To assist on-site inspection, each side might agree to label each missile with a tag that could not be counterfeited. Researchers at Lawrence Livermore and Sandia national labs are investigating several schemes for tagging mobile missiles (18). These tags, to be acceptable, cannot interfere with the missile's operation, and must not be able to act as a homing device. Among the ideas being explored are microchip tags that could be queried electronically, an acoustical hologram or a photomicrograph of a patch of the missile's skin, an imprint mold to be destroyed after a certain number of uses or a speckle pattern -- a random pattern of particles that could be read optically. Mitre Corp. has developed an electronic tag that can be read with public key encryption with a high degree of confidence against counterfeiting.

Inspectors might check all missiles within certain areas, chosen randomly, to assure that all missiles bear a tag. Missiles could be checked for tags as they entered and left the deployment area. In that way, if one nation built prohibited missiles, it would also have to build a separate infrastructure for maintenance and storage because the untagged missiles would be denied access to the designated deployment area.

The value of challenge inspections can be affected by the circumstances under which they are conducted. A report from Lawrence Livermore National Laboratory stated that the most effective challenge inspections feature a standdown (during which all traffic to, from and within a base ceases), occur with only one or two days notice and involve a large number of the possible bases (16).

Monitoring rail-mobile missile deployments

Accurate monitoring of the number of missile-carrying trains is more important than accurate monitoring of single-warhead missiles because, in the systems planned by the US and deployed by the USSR, a single train will carry several missiles and each missile will have multiple warheads.

The only deployed rail-mobile ICBM is the Soviet SS-24, and it is still in the early stages of deployment (19) The United States is developing a rail-mobile version of the MX missile. Rail-mobile missiles can be difficult to count by satellite reconnaissance, especially if the cars which carry them are not observably different from ordinary rail cars (Figures 4 and 5.) The Soviets have suggested that the SS-24 trains could be made to look different from ordinary rail cars so they could be observed by NTM (20).

The US plans to garrison its rail-mobile MX, and these garrisons are likely to have distinctive characteristics much like those of the SS-20 and SS-25 bases. At present, the Air Force plans for all MX rail garrisons to be on SAC bases. Although garrison basing facilitates verification, garrisoned missiles are in principle more vulnerable to strategic attack with short warning times, such as might come from submarine-launched ballistic missiles.

Figure 4. Artist's concept of a Soviet SS-24 missile being transported on a railroad car. (From *Soviet Military Power 1986*, US Department of Defense.)

Figure 5. Artist's concept of a Soviet rail-mobile SS-24 missile, at a maintenance facility. (From *Soviet Military Power 1988*, US Department of Defense.)

At the START negotiations, the United States and the Soviet Union reportedly have agreed that rail-mobile ICBMs will be confined to rail garrisons, each containing a limited number of missiles and launchers. Movements of missiles outside of the garrison would be allowed for routine movements and for dispersals, but prior notification would be required (21).

If rail mobile missiles were not to be garrison-based, they might be confined to designated rail networks -- the equivalent of designated deployment areas. Each side could then establish perimeter portal monitors at the nodal points of the designated rail network. However, restriction of rail-mobile missiles to specified networks would increase costs and decrease survivability.

Even if rail-mobile missiles were deployed on regular rail lines, their deployment mode and operations may be distinctive and observable. A missile-carrying train would need a certain number of cars for command and control, and for security and maintenance, but it is unlikely to pull any additional cars. The US plan for a rail-mobile version of the MX calls for each missile train to consist of six cars: two security cars, one maintenance car, one launch control car and two missile cars, each carrying one missile. The SS-24 (Figure 4) and the MX, each weighing about 100 tons, will require special, distinctive installations for loading missiles onto the rail tracks. Such installations could provide additional opportunity to monitor the number of deployed rail-mobile missiles (22).

Monitoring production facilities

The largest indivisible parts of a solid-fueled missile are the lower stages. Production facilities for missile stages -- especially for the solid-fueled mobile missiles -- have some distinctive characteristics that facilitate their identification (23).

The propellant used in solid-fueled rockets is an explosive. Despite safety regulations, fires and explosions are common in missile production facilities. Between 1984 and 1986, Morton Thiokol, the largest US manufacturer of solid-fueled rockets, experienced four major explosions and fires. In December, 1987, a fire in an MX missile casting facility killed four Morton Thiokol workers (24). Such accidents considered to be "an inherent risk in explosive propellant operations" (25).

The design, construction and operating procedures of solid-fuel production facilities reflect these risks. Because the propellants burn at about 6000 °F, explosive accidents are likely to ignite the entire building (26). Propellant is mixed far from the casting area, which again is located far from inhabited buildings. The result is that solid-fuel production facilities are large and consist of widely separated buildings, often protected by earth revetments.

The Bacchus West Works of Hercules Aerospace, while the most modern facility of its kind, exhibits the standard characteristics of a solid-fueled rocket production facility (Figure 6). These characteristics enable solid-fueled rocket production facilities to be identified by NTMs. Of course, Soviet production facilities are not built to US government standards, but the explosive nature of the materials used in rocket motor production places severe constraints on the design and operation of production facilities. The difficulty of manufacturing satisfactory solid-fueled rockets further constrains the feasibility of attempts at clandestine production (27).

Solid-fueled rocket production is also inherently slow. At Morton Thiokol, the time between the delivery of a rocket casing to the facility and shipping of a completed rocket is on the order of two months (28). In general, three to five days are required to cure the propellant after it has been cast (29). Lining and insulating the missile case also takes several days. Other processes -- such as inspections, painting and transportation within the plant -- increase the manufacturing time. As missile size increases, so do the handling difficulty and the manufacture time.

At present there is little information available on Soviet solid-fueled missile production facilities. But a study by the Congressional Research Service reported interviews with Morton Thiokol officials and US government officials as follows: "US detection of construction sites for the SS-X-24 and SS-25 is in part made possible because the propellants for such solid-fueled missiles must undergo a casting and curing process that - due to the danger of fire and explosion - occurs in a series of small buildings separated from one another. In the United States, such plants often cover hundreds of acres. Soviet plants for solid fuel missiles exhibit a similar pattern, though in a somewhat more confined space. Detection of such plants by US satellite photography has not proved difficult"

(30). START negotiators have considered banning liquid-fueled mobile missiles, since liquid-fueled missile production facilities may be easier to conceal (31).

Under a complete ban on mobile missiles, it would be possible to close facilities that were dedicated to the manufacture of these missiles, or to provide for on-site monitoring to ensure that these facilities did not manufacture the banned missiles. The identifiable characteristics of production facilities decreases the probability that a facility could produce solid-fueled missiles without being detected. However, on-site inspection may be required to ascertain what types of missiles were being produced at each facility.

Some solid-fueled rockets, such as space boosters and short-range missiles, are not likely to be restricted by any treaty in the near future. Because their manufacturing processes and operational activities are similar, unrestricted solid-fuel production could conceivably mask illicit production. One challenge of verification is thus to distinguish permitted from prohibited production. In the INF Treaty, the resemblance of the first stage of a permitted missile (SS-25) to

Figure 6. Bacchus West facility for solid rocket production. Covered passageways located on the right and connecting into the upper levels of buildings are for moving dry ingredients (top right). The left passageway connects the aluminum propellant premix building in the foreground with the two propellant mixing buildings, the prebatch and binder premix storage buildings, the mix bowl cleaning building, and the cast/cure building in the background. [Photo courtesy of the Hercules Aerospace Company.]

that of a banned missile (SS-20) prompted the provisions allowing the United States a continual presence at the SS-25 final assembly facility. Whether on-site inspection of mobile ICBM production facilities will be implemented under the START Treaty may depend in part on the degree of similarity between allowed and prohibited missiles. If the missile leaves the production plant in a canister, NTMs may be incapable of distinguishing the contents. If so, the parties to the treaty might want provisions for on-site inspection of these canisters, with use of cameras, x-ray equipment, etc.

It is also possible that a treaty might limit but not ban further missile production. In this case verification of the number of missiles produced would be required. Satellite monitoring can produce a rough estimate of the production rate, through observation of the number of mixers and casting facilities, number of workers, approximate number of shipments leaving a facility, and raw materials coming in. The US currently combines such data with other intelligence information to estimate Soviet missile production.

However, it is difficult to get an exact count of the number of missile stages produced using NTM only; photographic satellite coverage of a given spot is likely to be on the order of once a day at best. Missiles are primarily produced indoors and might not be shipped out when a satellite was overhead. And while solid-fueled missile production facilities can be identified with NTM, it is difficult to determine the type of solid-fueled missile being produced.

On-site inspection or cooperative verification measures could be used to count missile production. Measures could include both continuous monitoring at declared production facilities and periodic or challenge inspections at other production facilities and at missile assembly, maintenance and storage facilities.

A complication of production limitations is the practice of producing as many as twice the number of missiles actually deployed (32). The extra missiles are used for testing and replacement. Even though the extra missiles could be kept in monitored storage areas, if the treaty were broken these missiles could be made operational. To a great extent this practice has been adopted as an economy, to ensure that missile production lines will not have to be reopened in the future. Limits on the number of extra missiles produced would significantly reduce the risk of "breakout" from the treaty (33).

Monitoring other properties of missiles

Although most attention focuses on limiting the number of deployed missiles, the United States and the Soviet Union may want to limit other properties of ICBMs as well.

Concern about stability and survivability may motivate *limits on the number of warheads* on ICBMs. SALT II adopted a counting rule: Each missile was counted as having the same number of warheads as the maximum number of reentry vehicles with which it had been tested (34). For a possible restriction to single-warhead missiles, the same counting rule might be followed. Naturally, such a provision could limit only new types of missiles (35) to one warhead each (except for existing missiles that have been tested with only one warhead).

To verify that only one RV was deployed during a test, each side could observe the test shot with radar and intercept relevant telemetry. The agreement would have to provide for prior notification of such tests and ban encryption of necessary telemetry. Two possible problems present themselves, however. One is the possibility that the bus might maneuver as if it were dispensing more than one RV even if only one RV were actually launched. The Soviet SS-18 missile made 12 maneuvers during some of its tests, although it dispensed only 10 warheads. This may indicate that the SS-18 has the capability to dispense more than 10 warheads (36). If each side had confidence in the ability of its radars to sense these small maneuvers, it might ban all maneuvers but the one to release the single allowed warhead. The other possible problem is that a missile might be tested with a warhead that weighs less than the full throwweight of the missile, leaving open the possibility that the other side might deploy more than one warhead on that missile. SALT II does not allow tests in which the RVs weigh less than half the throwweight, although disputes have arisen over what constitutes the payload for a given missile.

Cooperative measures might be applied to this problem. Inspectors from one nation might be allowed to examine on-site the nose cone of a given missile of the other nation. There are several methods by which the number of MIRVs could be determined, including visual inspection, passive radiation detection, or by monitoring scattered radiation or induced fission (37).

Strategic stability might also be increased by *limiting improvements in missile accuracy*. One measure would be to ban tests of MaRVs - warheads that maneuver as they reenter the atmosphere in such a way as to guide themselves to their target (38). There is some concern that such systems may be able to target mobile missiles, if they could receive data from satellites during flight, or even if the position of the mobile missiles could be established shortly before the MaRVs were launched. Currently, however, the intelligence cycle time is not short enough to allow such updating.

A ban on tests of MaRVs would require the ability of the sensor to detect small shifts in the directions of the RV, and might also require that the nation conducting missile tests not encrypt data that might relate to the trajectory. A ban on MaRVs will not be essential for the survivability of mobile missiles until intelligence capabilities enable each side to update the positions of missiles as they move about in real time.

Military significance of violations

Although the Soviets have in the past stretched the interpretation of some arms control treaty provisions, they have in fact adhered to provisions requiring numerical limits. The construction of an early warning radar at the inland site of Krasnoyarsk, prohibited by the Anti-Ballistic Missile Treaty, is the strongest case for a treaty violation. Amid considerable criticism from the US, the Soviets have stopped construction of the Krasnoyarsk radar. However, there is no evidence of Soviets violations on a scale that one might term militarily significant.

The paradox of verification is frequently summarized by a quote from Amrom Katz: "We have never found anything that the Soviets have successfully hidden " (39). It's true we can never see everything. The question then becomes whether we will see everything of military significance. Former Secretary of Defense, Harold Brown, in testimony before the Senate Foreign Relations Committee in July, 1979, said:

"Throughout our years of monitoring Soviet forces, we have observed no grain elevators being loaded with ICBMs, no SLBM launchers being installed on fishing trawlers. One theoretical explanation is that the Soviets are doing an extraordinary job of concealment - with absolutely no breaches of security or inadvertent disclosure of what would be a substantial effort. The other, far more plausible interpretation is that Soviet clandestine deployment of strategic forces, if it exists at all, is at such a low level as to be strategically insignificant." (40)

Conclusion

The measures discussed here - national technical means, designated deployment areas, data exchanges, on-site monitoring, challenge inspections, cooperative measures to enhance satellite monitoring, missile production monitoring, flight test monitoring, on-site inspection of warhead numbers - provide an unprecedented variety of options for verifying limits on land-based missiles. While detection of a violation cannot be guaranteed, a combination of these measures would complicate an attempt to evade the treaty, and would increase the probability that evasion would be detected.

Appendix A: National technical means of verification

Both the US and the USSR have deployed a wide range of sensors for "viewing" from afar the military developments of one another. The same sensors used to keep track of weapons developments can be used to monitor compliance with arms control treaties. National technical means are those technical measures which one nation may use to observe the military activities of the other without intruding on its sovereign territory. NTMs generally consist of sensors that monitor some portion of the electromagnetic spectrum. We can roughly categorize these technologies by the portion of the electromagnetic spectrum they employ; greater resolution is possible with radiation of shorter wavelengths. We will consider these technologies in order of increasing wavelength.

Photography. High-resolution cameras are mounted on satellites that circle the earth in polar orbits at distances from about 200 to 400 km above Earth's surface. The US currently relies heavily on satellites that can transmit digital images in real time, to provide surveillance with a resolution estimated at about 6 in (15 cm) (41). The US is scheduled soon to boost into orbit the first of a new generation of satellites, which are reported to have cameras with resolutions better than 10 cm, and to carry a thermal infrared camera as well (42,43).

One technique that has greatly improved the capabilities of satellite photography is the use of charge-coupled devices (44), which are semiconductors that store charge in linear proportion to the light exposure. These sensors have a greater dynamic range than film and enable real-time digital transmission of data, eliminating the necessity of retrieving film canisters ejected from the satellites. Techniques of digital image processing have further enhanced the photographic images collected.

Some idea of the capabilities of such photographic reconnaissance was revealed when Jane's Defense Weekly published a photograph taken by a US spy satellite of the construction of an aircraft carrier in a Soviet shipyard (45). The photograph shows many fine details of the vessel.

Infrared photography. Although photographs based on visible light provide excellent resolution, they can only be taken during the day. Thermal infrared imagery can image objects on earth by sensing the thermal radiation they emit. Infrared (IR) sensors can use thermal emissions for such tasks as distinguishing camouflage from natural vegetation, locating underground missile silos, which are warmer than their surroundings, or detecting heat from moving vehicles. The atmosphere is relatively transparent to IR radiation in two bands -- 3 to 5 microns and 8 to 14 microns. Because these IR wavelengths are about a factor of 10 longer than radiation in the visible region, the resolution possible with IR photographs is about a factor of ten worse. Very good IR sensors might have a temperature resolution of approximately 0.01 ^{o}K at close range and spatial resolution on the order of a few meters from orbit (39).

Radar Imagery. Neither visible nor IR photography can monitor activities on the ground when clouds obscure the landscape, as they commonly do in Western USSR. However, radar waves can penetrate the clouds. The antennae for radars must be quite large, although a technique called "synthetic aperture radar" artificially creates an enlarged antenna for radars on moving platforms such as airplanes and satellites. The US has built large ground-based radars to give strategic warning of an attack, and it uses one ground-based and one ship-based radar to monitor the return to Earth of reentry vehicles from Soviet missile flight tests. As an example of the capacity of such radar, the ground-based phased-array radar known as Cobra Dane, which monitors Soviet missile flight tests from Alaska, uses an array of about 35,000 small radiating and detecting elements simultaneously to track up to 100 objects, each the size of a grapefruit, at a distance of the order of 2000 km (46).

The resolution of radar images is on the order of tens of meters. For example, a synthetic aperture radar with an antenna length of 11 m on a satellite such as the US civilian Seasat satellite can get a resolution of about 4.9 meters at an orbital height of 800 km (47). The resolution could be improved to about 1 meter by using a carrier frequency as high as 10 GHz (47).

The US does not now use any radar satellites for strategic reconnaissance, but at least one satellite is reported to be under development for a launching in the late 1980s (42,43). The Soviets operate a space-based ocean reconnaissance satellite that uses active radar to detect and identify surface ships.

Sigint. Satellite-based sensors can also intercept radio waves and other electronic signals used for communications. The collection of such data is called "sigint," for signal intelligence. These data have been useful in providing clues to certain activities, warning of pending weapons tests and relaying by way of telemetry the results of such tests. Two sets of American Rhyolite satellites currently monitor tests at Plesetsk and Tyuratam in the USSR. They can "hear" signals as weak as 10 W with a 20-meter diameter antenna (48). Two other US satellites -- Magnum and Chalet -- also gather radio signals and telemetry.

Data interpretation. All these sensors gather such incredible amounts of data that they cannot be evaluated in a timely fashion. Priority is given to geographical regions where significant activity is expected, such as missile test centers, command posts and deployment areas. In these areas, the US expects that it will notice promptly any important changes. Digital Image Processing computers automatically spot changes from one image to another, but it requires more finesse to determine which changes are significant (43). Military developments in unforeseen regions may take longer to spot. Thus, for example, it took the US about one year to discover that the Soviets were constructing at the remote Siberian location of Krasnoyarsk a large, phased array radar that is probably a violation of the ABM Treaty.

Humint. Human intelligence, more crudely known as spying, remains a valuable component of one nation's assessment of the military activities of other nations. In the case of data interpretation, human intelligence can provide clues as to what segments of the collected data might yield the most significant information, and give direction to future collection of intelligence data.

Appendix B: On-site inspection provisions of the INF Treaty

On-site inspection provisions of the INF Treaty include baseline inspections, close-out inspections, elimination inspections, short-notice inspections and continuous portal monitoring.

The baseline inspections are for verification of data provided on numbers and locations of INF missiles, launchers and support equipment, and are to be conducted between 30 and 90 days after the Treaty comes into force. Inspections are restricted to specified facilities, which are located in Belgium, West Germany, Italy, the Netherlands, Britain, East Germany, Czechoslovakia, the United States, and the Soviet Union.

Close-out inspections are to verify that an INF facility no longer contains INF missiles or supports INF activities.

The United States and the Soviet Union are required to observe destruction of INF missiles at the elimination facilities (Treaty, Article XI, paragraph 7). Notification of missile elimination must be given at least 30 days in advance, or 10 days in advance if the missile is to be destroyed by launching (Treaty, Article IX).

Short-notice inspections are allowed at missile operating bases and support facilities for 13 years. 20 inspections per year are allowed for the first three years, 15 per year for the next five years, and 10 per year for the last 5 years (Treaty, Article XI, paragraph 5). An inspection must be announced by the inspecting side at least 16 hours before the inspectors arrive in the country. Between 4 and 24 hours after arrival, the inspecting side must identify the site to be inspected (Protocol on Inspection, Article IV, paragraph 2). Within 9 hours after this announcement, the inspected side must transport the inspectors to the specified site (Protocol on Inspections, Article VII). Inspectors may inspect the entire site, including the interior of structures, containers, or vehicles, provided that they are larger than the specified dimensions of INF missile stages (Protocol on Inspections, Article VII). Inspectors may bring cameras, portable weighing devices, radiation detection devices and other equipment. With the exception of cameras, all equipment may be operated by the inspecting team. Cameras must be operated by the inspected party at the request of the inspectors, and must be capable of producing duplicate, instant development prints. Each party will keep one copy of each photograph (Protocol on Inspections, Article VI, paragraph 9).

Continuous portal monitoring will be conducted at the SS-20 final assembly facility at Votkinsk and at the Pershing II production facility at Hercules Aerospace in Magna, Utah (Treaty, Article XI, paragraph 6). Monitoring will be allowed for up to thirteen years; if, after two years into the treaty, there is a 12 month period in which the Votkinsk facility does not assemble any missile stages which are outwardly similar to an INF missile (i.e., the SS-25), then the monitoring rights will end, for both parties, unless SS-25 assembly is initiated again (Treaty, Article XI, paragraph 6). (Note that the first stage of the SS-25 missile is outwardly similar to the first stage of the SS-20; Votkinsk is the final assembly facility for both missiles. The Pershing II did not have a final assembly facility; it was assembled at the deployment site.)

At the perimeter of the production or assembly facility, inspectors may install systems to monitor the exits, including weight sensors, surveillance systems and vehicle dimensional measuring equipment. At the portal, equipment for measuring length and diameter of missile stages, and non-damaging imaging equipment (such as X-ray equipment) may be installed (Protocol on Inspections, Article IX).

References and notes

* This work was done largely while Barbara Levi worked at the Center for Environmental Studies at Princeton University.

1. For the Treaty and Protocols, see Arms Control Today, Jan/Feb 1988.

2. S. Graybeal and M. Krepon, "The limitations of on-site inspection," Bulletin of Atomic Scientists, Dec 1987, p. 22.

3. "New spy satellites urged for verification," Science, 1 Apr 1988, p. 20.

4. Tom Cochran, "The NRDC/Soviet Academy of Sciences joint nuclear test ban verification project," Physics and Society, published by the American Physical Society's Forum on Physics and Society, Jul 1987, p. 5.

5. US State Department, "Gist", Dec 1987.

6. This plant is the Votkinsk Machine Building Plant, Udmurt Autonomous Soviet Socialist Republic, Russian Soviet Federative Socialist Republic.

7. James A. Schear, "Cooperative measures of verification: how necessary? how effective?", in *Verification and Arms Control*, Edited by William C. Potter, Lexington Books (1984), p. 21.

8. J. Adams, "Verification: peacekeeping by technical means," IEEE Spectrum, Jul 1986, p. 42.

9. *Legislative Environmental Impact Statement: Small Intercontinental Ballistic Missile Program*, AD-A173 827, US Air Force, Nov 1986.

10. This estimate of roughly 20 men per missile seems low in comparison with the 35 people per missile reportedly required to field the Pershing II in Germany. See J. Medalia, *Small Single-Warhead Intercontinental Ballistic Missiles: Hardware, Issues and Policy Choices*, CRS Report No. 83-106F, Congressional Research Service, Washington, D. C. (1983).

11. The weight of the launcher for the US Midgetman is twice the maximum highway load of 100,000 lb permitted in the US, and about four times heavier than a large tractor-trailer. The HML is over six times the weight of its missile, no doubt reflecting the requirement that it survive overpressures of 30 - 50 psi. A rule of thumb for trucks is that their weight is about twice that of the load they are transporting. The Soviet SS-25 is said to be about the size of the US Minuteman, whose weight is 76,000 lbs. If its transporter-erector-launcher is not hardened, the vehicle probably weighs about 160,000 lbs.

12. Michael Gordon, "New Data Reduce Russian Missile Force," New York Times, 18 Dec 1987.

13. Report of the Committee on Foreign Relations, *The INF Treaty*, US Senate, 14 Apr 1988, pp. 56-57. Also Michael Gordon, "Arms advisor says US can spot missile breach," New York Times, 29 Jan 1988, p. 4.

14. Paul Nitze's testimony, Hearings on the INF Treaty, US Senate Committee on Foreign Relations, 28 Jan 1988, Part 1, p. 293.

15. The Arms Control Reporter, 611.B.481, 1988.

16. F. Holzer, "On-site inspection: verifying the inventory of mobile, intermediate-range missiles," Engineering and Technology Review, Lawrence Livermore National Laboratory, Aug 1986, p. 20-25.

17. INF Treaty, Article XII, paragraph 3. See Reference 1.

18. F. Holzer, op. cit. See also IEEE Spectrum, Jul 1986, p. 55.

19. The New York Times, 9 Aug 1987, p. 3.

20. The New York Times, 22 Jun 1986, p. 14.

21. The Arms Control Reporter, 1988, p. 611.D.76, summary from US START delegation, 1 Jun 1988.

22. Peter D. Zimmerman, "A good place for missiles is an obvious one: on trains," Los Angeles Time, 13 Aug 1987, Part II, p. 7.

23. Much of the discussion in this section has been published in: Valerie Thomas, "Monitoring solid-fueled missile production for arms control," Physics and Society, Jan 1988.

24. The New York Times, 30 Dec 1987, p. 1. USAF Mishap Report, Air Force Plant 78, Morton Thiokol Inc Strategic Division, 29 Dec 1987.

25. US General Accounting Office, "Space shuttle: NASA's procurement of solid rocket booster motors," GAO/NSIAFD-86-194, US Government Printing Office, Washington, D.C. (1986), pp. 12-13.

26. George P. Sutton, ***Rocket Propulsion Elements***, 5th edition, John Wiley and Sons, New York (1986), p. 293.

27. These identifiable features are characteristic only of solid-fueled rocket production facilities. Final assembly plants, such as the Votkinsk plant, or facilities that produce liquid-fueled rockets, may be more difficult to identify.

28. Private communication, Roger Boisjoly, former engineer, Morton Thiokol, Inc., 8 Nov 1987.

29. Seymour M. Kaye, *Encyclopedia of Explosives and Related Items*, p. P412.

30. Stuart D. Goldman, Paul E. Gallis, and Jeanette M. Voas, "Verifying arms control agreements: the Soviet view," Congressional Research Service Report CRS-87-316F, 1987, p. 82. START negotiators have considered banning liquid-fueled mobile missiles, since liquid-fueled missile production facilities may be easier to conceal.

31. The Arms Control Reporter, 1988, p. 611.D.76, from US delegation summary, 1 Jun 1988.

32. Thomas B. Cochran, William M. Arkin, and Milton M. Hoenig, *Nuclear Weapons Databook, Vol. I: US Nuclear Forces and Capabilities*, Ballinger, Cambridge (1984), pp. 117, 121, 137, 142, and 145.

33. "Breakout, verification, and force structure: dealing with the full implications of START," House Armed Services Committee, Report of the Defense Policy Panel, 24 May 1988.

34. SALT II Treaty, Article IV, paragraph 10.

35. A new type of missile is one that can be distinguished from existing types by national technical means.

36. Herbert Scoville Jr., "Verification of Soviet Strategic Missile Tests," in *Verification and SALT: The Challenge of Strategic Deception*, edited by William C. Potter, Westview Press, Boulder (1980).

37. Robert Mozley, "Reducing the number of multiple warhead missiles," in *Verification of Nuclear Warhead Reductions and Space-Reactor Limitations*, edited by Frank von Hippel and Roald Sagdeev, Gordon and Breach, New York (forthcoming).

38. Matthew Bunn, "The next nuclear offensive," Technology Review, Jan 1988.

39. Amrom Katz, "The fabric of verification: the warp and the woof," in *Verification and SALT: The Challenge of Strategic Deception*, edited by William C. Potter, Westview Press, Boulder (1980), p. 212.

40. H. Brown, quoted in IEEE Spectrum, Jul 1986, p. 54.

41. J. Richelson, Arms Control Today, Oct 1986, p. 15.

42. John Pike, "Satellite reconnaissance," Harvard International Review, Aug/Sep 1988.

43. J. A. Adam, "Counting the Weapons," IEEE Spectrum, Jul 1986, p. 49.

44. See for example David Hafemeister, "Emerging technologies for verification of arms control treaties," American Journal of Physics, Volume 54, 1986, pp. 693-699.

45. Aviation Week and Space Technology, 13 Aug 1984, pp. 26-27.

46. E. Brookner, Scientific American, Feb 1985, p. 94.

47. E. Brookner, "Radar imaging for arms control," in *Arms Control Verification: The Technologies That Make It Possible*, K. Tsipis, D. W. Hafemeister, and P. Janeway, editors., Pergamon-Brassey's, Washington D.C. (1986), p. 139-141.

48. J. Richelson, "Monitoring the Soviet Military," Arms Control Today, p. 14, October, 1986.

Part V Chapter 3

Minuteman/MX system: becoming vulnerable?

Art Hobson

Introduction

Is there actually a vulnerability problem with US land-based missiles? If so, what are its dimensions, today and tomorrow? Its implications?

This Chapter presents the "standard" theory of silo destruction, and analyzes the vulnerability of the Minuteman (MM)/MX system (see Figure 1) to plausible Soviet attacks today and in the mid-1900s, a likely time-frame for new ICBM deployments.

Obviously, the conclusions will not be certain. Uncertainties arise not from the theory, which is generally accepted by most analysts, but from the parameter inserted into the theory when making specific predictions. Unfortunately, uncertainties are not always acknowledged in the literature. Here, I include the complete range of plausible values of all parameters to calculate a range of plausible silo vulnerabilities.

The main conclusion is that the MM/MX system might still be survivable today, but if present trends in Soviet missile accuracy continue then it will not be survivable tomorrow (mid-90s). For a more precise and quantitative summary of the main findings of this Chapter, see Figure 4 below.

Figure 1. MX misssile in modified Minuteman silo. [US Air Force drawing.]

The standard model of silo vulnerability

This section summarizes the standard model of silo vulnerability. Most analysts agree that the model gives reasonable predictions and that any improvements on the assumptions of the theory would alter the predictions by only a small percentage. In most cases, such improvements have negligible strategic or policy implications. In any case, inaccuracies in the theory are far overshadowed by the large error bars in the data that is put into the theory to make quantitative predictions.

The point of impact of the attacking warhead is assumed to have a 2-dimensional Gaussian probability distribution whose density is

$$f(\vec{r}) = \frac{1}{2\pi\sigma^2} \exp\left\{\frac{-\vec{r}\cdot\vec{r}}{2\sigma^2}\right\} \qquad (1)$$

where **r** is the vector in the plane of the impact from the target silo to the point of impact. An elliptically-symmetric distribution would be more accurate than the circular symmetry of Eq. (1) but would not significantly alter our conclusions. The source of the statistical dispersion σ is the random targeting error arising from a wide variety of sources (1,2). Sometimes **r** is replaced by **r**-**B** in Eq. (1), where **B** is the systematic error or "bias" caused by such effects as magnetic forces over the north pole. Such systematic errors move the center of the Gaussian distribution to some point **B** relative to the target. It is usually assumed that bias errors are negligible, although there has been some controversy on this point (3-6). We will assume that the bias is negligible, i.e. $B \ll \sigma$.

As a measure of the inaccuracy of the attack, the "CEP" is usually used instead of σ. The CEP is the radius of that circle (the "circle of equal probability") centered on the target such that Prob (r<CEP) = 0.5. Integrating Eq. (1) from r = 0 to r = CEP, and setting the integral equal to 0.5, one finds

$$\exp(-CEP^2/2\sigma^2) = 0.5 \qquad (2)$$

from which CEP = 1.18σ.

Let us assume that the warhead has a specific radius of destruction RD against a specific type of silo, i.e. a fixed distance from the silo within which the silo will surely be destroyed and outside of which it will surely survive. (It would be more accurate to replace this "cookie cutter" assumption with a continuous probability distribution that drops rapidly from a constant inside the circle of equal probability to zero just outside, but again this would change our conclusions only slightly.) Integrating Eq. (1) over (RD<r<infinity) and setting the integral equal to SSPS, one obtains the "single shot probability of survival"

$$SSPS = \exp(-RD^2/2\sigma^2) = 0.5^x \quad \text{where } x = RD^2/CEP^2 \ , \qquad (3)$$

where Eq. (2) was used in making the transition from base e to base 0.5. Equation (3) is graphed in Figure 2. As a check on Eq. (3), note that SSPS = 0.5 when CEP = RD.

RD depends on the blast, and on the silo. Silos are vulnerable to "ground shock" waves passing directly through the ground, "airblast overpressure" or "blast" shock waves in the air, heating, and nuclear effects such as x-rays, gamma-rays, electromagnetic pulse, and neutrons. Silos are designed to withstand blast up to a certain threshold, and to withstand all other effects that are likely to accompany blasts up to this level. Thus, we assume that blast is the dominant destruction mechanism.

Experimentally, the airblast overpressure (in atmospheres above atmospheric pressure) from a groundburst of yield y (in megatons, MT) at a distance r (in kilometers) from the blast is (1,2)

$$P = 6.31\ y/r^3 + 2.20\ \sqrt{y/r3} \approx 7.04\ y/r^3 \qquad (4)$$

where the approximation is valid for larger overpressures, in the range 5-500 atm. The experimental data supporting Eq. (4) may be found in (7).

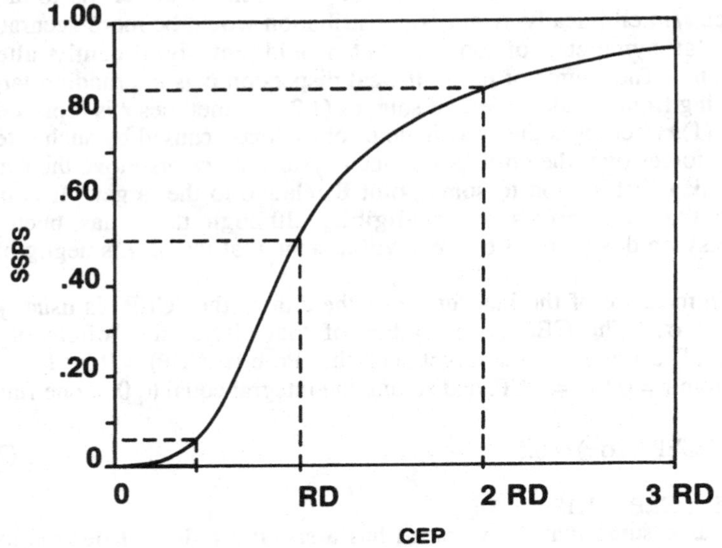

Figure 2. Single shot probability of survival (SSPS) of a point target as a function of the warhead's CEP, Eq. (3).

The maximum airblast overpressure a silo can withstand is called its "strength" S. Thus P = S when r = RD. Solving Eq. (4),

$$\sqrt{y/_{RD^3}} = \frac{\{5.21\ S + 1\}^{1/2} - 1}{5.74} \approx 0.398\ \sqrt{S} \qquad (5)$$

where the approximation is valid for silos (S> 100 atm) but not for softer targets. Solving the approximation for RD and plugging the results into Eq. (3),

$$\text{SSPS} = 0.5^x, \text{ where } x = 3.42\ L/S^{2/3}, \qquad (6)$$

where S is the silo strength in atm and L is the warhead's "lethality" defined by

$$L = y^{2/3}/\text{CEP}^2 \qquad (7)$$

where y is the yield in megatons and CEP is in km. Equation (6) expresses the silo's survival probability in terms of the missile and silo "figures of merit" L and S. Figure 3 is a graph of SSPS, Eq. (6), as a function of L and S. Since L and S are graphed logarithmically, the contours of equal SSPS are straight lines. The approximate lethalities of several present and future missile warheads are

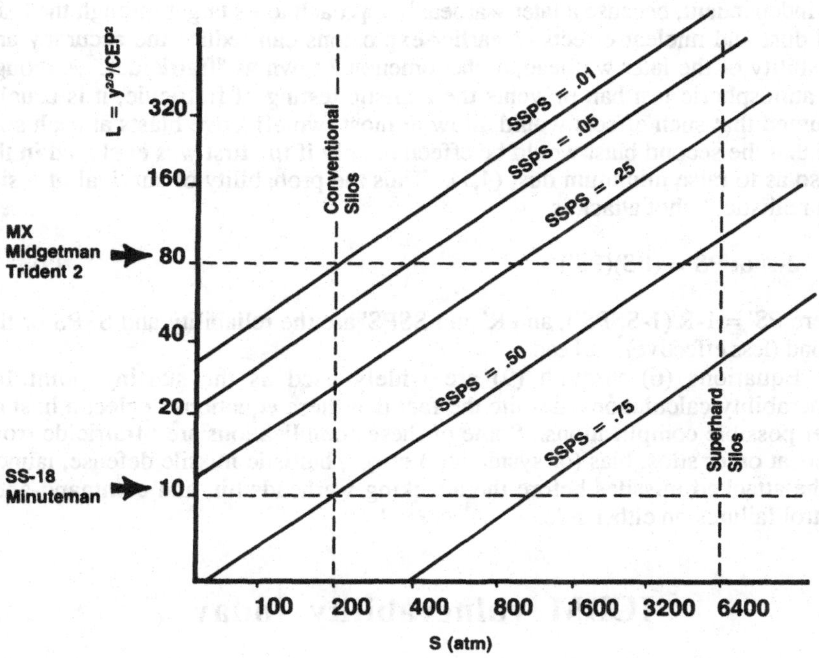

Figure 3. Single shot probability of survival for silos of strength S attacked by warheads of lethality L, Eq. (6).

shown, as are the approximate strengths of conventional and superhard (Chapter 11) silos. As examples, note that conventional silos have a survival probability of about 50% against SS-18 or Minuteman-3A warheads, and a survival probability of only 1% against MX warheads.

Equation (6) is an idealized approximation. Many additional factors will come into play in a real attack. Some of these are quantitatively analyzable. Of the analyzable factors, three stand out: the attacking warhead might fail; more than one warhead might be directed at a single silo; and the effects from one exploding warhead might destroy or reduce the effectiveness of other warheads.

The probability that a given warhead-type will not fail, i.e. that it will be boosted and will deploy and re-enter and explode as designed, is known as the warhead's "reliability" R. Taking reliability into account, the probability of survival of a single silo against a single warhead becomes

$$PS = 1 - PD = 1 - R(1\text{-}SSPS) \quad (8)$$

since the probability of destruction is PD = R(1-SSPS), where SSPS is given by Eq. (6).

If N warheads attack the same silo, and if their effects are statistically independent, the overall survival probability will be $(PS)^N$. But the effects are not independent, because a later warhead's approach to its target through the blast and dust and nuclear effects of earlier explosions can reduce the accuracy and reliability of the later warhead, a phenomenon known as "fratricide". Although the atmospheric test ban prevents the realistic testing of fratricide, it is usually assumed that such effects would allow at most two effective blasts at each silo, and that the second blast would be effective only if the first was exploded in the air so as to raise minimum dust (4,7). Thus the probability of survival of a silo in a realistic 2-shot attack is

$$\text{2-shot PS} = (PS)(PS') \qquad (9)$$

where PS' = 1-R'(1-SSPS'), and R' and SSPS' are the reliability and SSPS of the second (less effective) warhead.

Equations (6) through (9) are widely used as the starting point for vulnerability calculations, despite the fact that these equations neglect a host of other possible complications. Some of these complications are : fratricide from blasts at other silos, bias (or systematic) errors, ballistic missile defense, launch of the attacked missiles before the attacking warheads hit, and command-and-control failures on either side.

ICBM vulnerability today

Claims about US ICBM vulnerability range from predictions that an attack by some 200 Soviet SS-18 missiles would destroy 90% of the 1000 US ICBM silos (9), to predictions of at little as 50% destruction (3,4). Many congressional and DOD sources estimate ICBM vulnerability in the 80-90% range (10-11) or in the 65-80% range (12,13). None of the congressional and DOD sources give details of the supporting calculations. Political biases might account for some of the disagreement: "hawks" lean toward worst-case arguments showing that old weapons (e.g. Minuteman) are vulnerable and need to be modernized, while "doves" lean toward best-case arguments showing that the old weapons still suffice. In analyzing proposed *future* weapons systems (e.g. Midgetman, SDI), on the other hand, hawks favor best-case arguments showing that the proposed new system will be highly effective, while doves favor worst-case arguments showing that the proposed weapons are not worth building.

Technically, the disagreement stems from a lack of consensus about the values of the parameters in Eqs. (6) - (9). Indeed, there may be no single correct value. For example, a warhead's CEP may vary with weather conditions at the target, and its reliability may vary with its age.

Our approach will be to calculate a plausible *range* of vulnerabilities, or "vulnerability error bars," corresponding to the range of plausible values of the parameters. Thus we assume pessimistic (for US ICBM survivability) but plausible values of the parameters, and then we assume optimistic but plausible values. We use the two sets of parameters to make a worst-case and best-case estimate of US ICBM vulnerability, representing the extremes of the vulnerability error bars.

Today, any attack on MM/MX silos would be carried out by Soviet ICBMs, since Soviet SLBMs lack the accuracy needed for such an attack. Furthermore, the 10-warhead SS-18 would probably be used since the other plausible silo-based candidates have little silo-destruction capability (12), while any presently-deployed mobile SS-24s and SS-25s would probably be held in reserve due to their greater survivability.

A 2-on-1 attack by SS-18s against the 1000 MM/MX silos would require 2000 warheads. This is 65% of the Soviet SS-18 force, or 30% of the Soviet ICBM warhead inventory, or 20% of all Soviet strategic warheads. How effective would such an attack be?

Plausible values of the SS-18s CEP range from the 250-300 m quoted a few years ago in the open literature (14) to the 200-250 m generally assumed at DOD today. We will use 200-250 m as our range of plausible estimates since the upper range of 300 m quoted a few years ago is reduced by now due to recent accuracy improvements (12,13). From Eq. (9), the SS-18 lethality is then L = 10-16. MM/MX silo strength is usually given as S = 133 atm (2000 psi) (15), but it might be as high as 200 atm because new studies of silo hardening have revised official estimates of the strength of existing US and Soviet silos upward by about 50% (16). The strength of the 50 MX silos is about the same as that of Minuteman silos, since MX silos are just refurbished Minuteman silos with little upgrading except for a new shock isolation system (17). Putting these estimates into Eq. (6) leads to SSPS = 23% at the pessimistic end, and to SSPS = 50% at the optimistic end. Since this implies that at least 230-500 of the existing 1000 US silos would survive a 1-on-1 attack, the Soviets would probably try to explode a second warhead at each silo.

What about reliability and fratricide? Estimates of overall US or Soviet warhead reliability (i.e. probability of successful launch and flight and explosion) fall in the range 70-90% (4,18-22). But this is not the whole story, because most failures would occur in the boost and post-boost stages, and it may be possible to launch replacements at least for those missiles observed to fail in these early stages (23-25). This could be done, for example, by maintaining a replacement set of missiles pre-programmed for the missions of the attacking missiles, any of which could be launched immediately upon detection of an early failure. More specific re-programming of reserve missiles for the missions of those missiles or warheads observed to fail is difficult and would probably take an hour or more, which might be too long to be militarily useful.

Not much is known about fratricide, because the atmospheric test ban prevents realistic tests of the effects of nuclear explosions on other incoming warheads. Indeed, I have found very few discussions that provide any quantitative detail. A pessimistic analysis should assume that in an initial airburst (to minimize effects on the second incoming warhead) with a followup groundburst, fratricide effects are essentially zero. One quantitative optimistic analysis (3,4) assumes that 5% of the second wave is destroyed by fratricide, and that the remainder of the second wave have a doubled reentry error (the portion of the inaccuracy that is ascribable to atmospheric buffeting during reentry).

Making the pessimistic assumptions that R=0.90 with no fratricide, and combining this with the pessimistic estimate that SSPS=23%, the overall two-shot probability of survival Eq. (9) is 9%.

For the optimistic assessment, assume R=0.70 in the first wave. Fratricide then further reduces the effectiveness of the second wave. Assuming, rather arbitrarily but as in other optimistic analyses (3,4), that 5% of the second wave is destroyed by fratricide, and that the remainder have a doubled reentry error, the second wave's reliability becomes R'=0.65. The second wave's CEP is increased above the 250 m CEP assumed for the first wave and is calculated as follows: A normal ICBM warhead reentry error is 90 m (3). For a 250 m CEP, the *non*-reentry error x is thus found from

$$(x^2 + 90^2)^{1/2} = 250,$$

from which x = 233 m. Doubling the 90 m reentry error,

$$CEP' = (233^2 + 180^2)^{1/2} = 295 \text{ m}.$$

Putting these optimistic values into Eqs. (6) - (9), the 2-shot survival probability is 49%.

These results are graphed in Figure 4 and tabulated, along with the assumed values of the parameters, in Table 1. In addition to the two extreme cases calculated above, Figure 4 shows one case with assumptions intermediate to the pessimistic and optimistic cases; Table 1 lists these intermediate assumptions.

Figure 4 shows the number of US MM/MX silos surviving a Soviet SS-18 ICBM attack today, out of the 1000 silos actually deployed. Note that even in the most pessimistic case, there are 90 surviving silos, representing some 210 warheads (since an expected 5% of the survivors would be 10-warhead MXs and the remainder would be MM with an average of 2 warheads/silo). This is an enormous retaliatory force. For example, it could destroy 210 Soviet cities. More importantly, the bomber and submarine legs of the US triad would contain very large surviving forces.

Table 1. Survivability of the MM/MX force today, to a 2-wave attack by Soviet SS-18 ICBMs. Assumptions and conclusions.

	pessimistic assumptions	intermediate assumptions	optimistic assumptions
CEP	200 m	225 m	250 m
SS-18 yield	0.5 MT	0.5 MT	0.5 MT
MM/MX silo strength	130 atm	165 atm	200 atm
SSPS	23%	38%	50%
Reliability*	90%	80%	70%
Fratricide	0	3%	5%
Fraction surviving	9%	27%	49%

*including possible replacements.

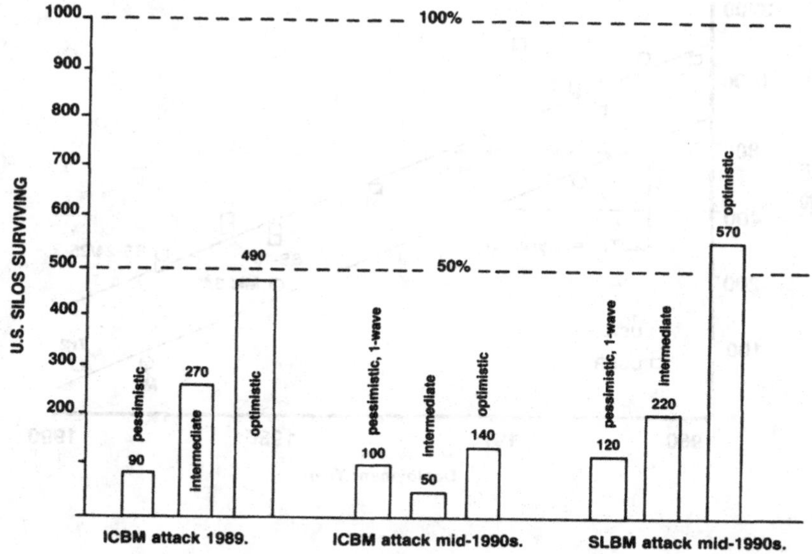

Figure 4. Survivability of the 1000 MM/MX force under Soviet attack, today and in the mid-1990s. Except for the two cases noted, all attacks are 2-wave. Pessimistic assumptions are: low CEP, 2000 psi silo strength, reliability=90%. Optimistic assumptions are: high CEP, 3000 psi silo strength, reliability=70%. For more detail, see Tables 1, 2, and 3.

The future: attack by ICBMs

How vulnerable would the present MM/MX force be to a Soviet attack in the mid-1990s? During the next decade, Soviet capabilities against US ICBMs are likely to have improved in at least two important ways: ICBM accuracies will improve, and submarine-launched ballistic missile (SLBM) accuracies may have improved to the point that SLBMs will be able to destroy US silos. We study ICBM and SLBM attacks separately.

First, ICBMs. We estimate future Soviet ICBM capabilities in two ways: from historical trends (26), and from plausible Soviet technological accomplishments during the next decade. Historically, since the first missile deployments around 1960, the accuracy of both US and Soviet missiles has doubled every 7 years, i.e. CEPs have halved every 7 years. In addition, Soviet accuracy improvements have lagged US accuracy improvements by 7 years. Both the 7-year halving time and the 7-year lag time are shown in Figure 5 (26).

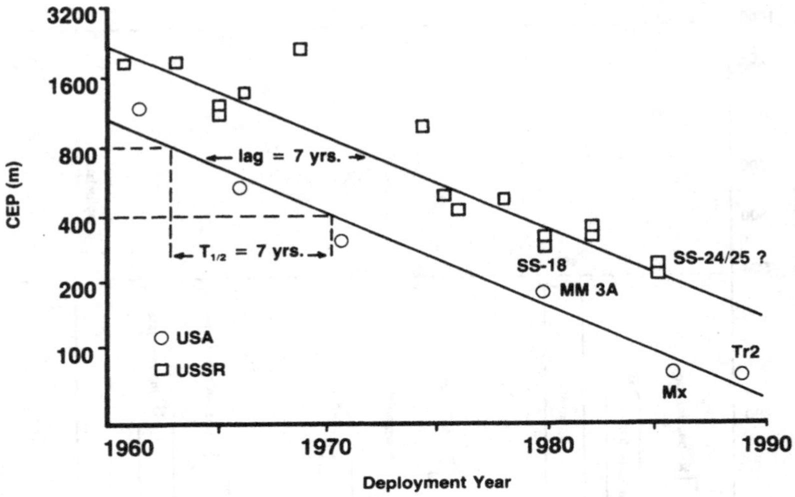

Figure 5. US and Soviet ICBM and SLBM accuracies, historically. Graph based on D. Schroeer, *Science, Technology, and the Arms Race* (Wiley, New York, 1984).

Since the most accurate Soviet ICBMs today have CEPs of 200-250 m, the 7-year lag time implies a mid-90s CEP of 100-125 m. On the other hand, today's most accurate US ICBM, the MX, has a 90 m CEP (27), so the 7-year lag implies a future Soviet CEP of 90 m.

From a technical perspective, the MX's very small CEP is achieved through the ultimate in inertial guidance (Figure 6). It would be difficult to do better without terminal guidance, because ICBM re-entry errors alone are about 90 m (3). That is, the MX guidance system has eliminated essentially all of the launch- and boost-phase errors. It is not expected that the Soviets will be able to deploy terminal guidance by the mid-90s (20), a concept that is only now being considered for future US ICBMs and SLBMs (15,20,23,28). Thus 90 m is the best the Soviets could hope to achieve by the mid-90s. On the other hand they may fall short of this goal, because their inertial guidance may be less than perfect. Thus a CEP in the 90-125 m range, estimated above via historical trends, is plausible on technical grounds as well. Further support is found in (12): "Test flights have now begun for the follow-on missile to the SS-18 Mod 4, a missile that will probably have greater throw-weight, carry at least 10 warheads, and have greater accuracy---increasing its effectiveness against US ICBM silos ---."

At the pessimistic end, we make all the worst-case assumptions of the preceding section, but with the 200 m CEP replaced by 90 m. This small CEP implies SSPS = .001 = 0.1%, according to Eq. (6). Since the worst-case Soviet

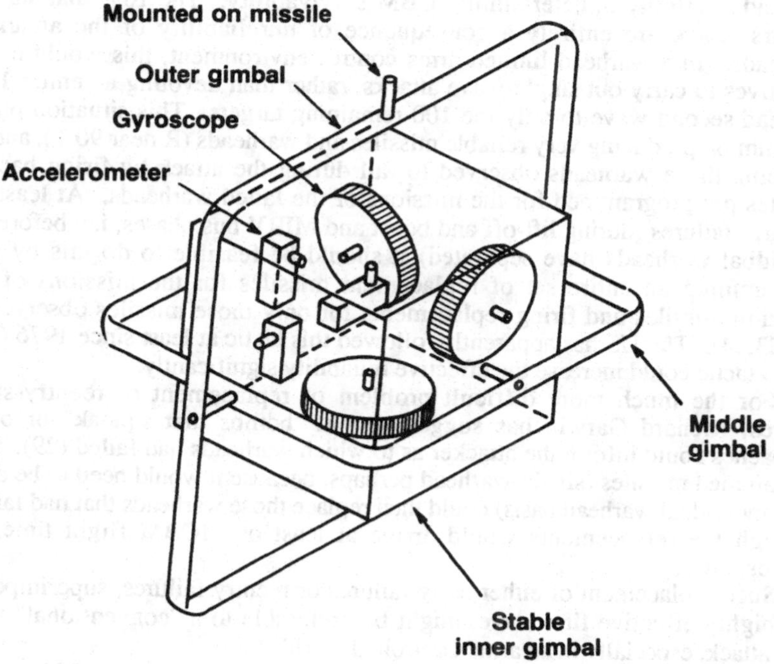

Figure 6. Inertial guidance. The inner platform, containing three accelerometers, maintains a fixed spatial orientation regardless of the missile's motion. This orientation is maintained by the three freely rotating gimbals and by the three high-frequency gyroscopes mounted to the inner platform. A large torque would be needed to change the orientation of any one of the three mutually perpendicular gyroscopes. The acceleration in each direction is measured by the deflection of the accelerometer associated with that direction, just as a pendulum hanging from the ceiling of a moving vehicle measures the vehicle's acceleration. Computers then integrate the acceleration to obtain the velocity and position, relative to the initial velocity and position, at any instant during the flight.

reliability is R = 90%, we have a survival probability in a single-wave attack of only 10%. The overall two-wave probability of survival is just 1%. However, if the Soviets believed something like these pessimistic (for the US) assumptions, they might choose not to launch a second wave. This has the important effect of allowing the attack to be carried out with a smaller 1000 warhead force, saving the 1000 second-wave warheads for other purposes.

Especially in an arms-controlled world, with warhead limits on both sides, this could be an important consideration.

Another notable feature of this worst-case estimate is the dominance of warhead reliability in determining ICBM survivability: The 10% that survive the first wave are entirely a consequence of unreliability of the attacking warheads. In a warhead-limited arms control environment, this would create incentives to carry out single-wave attacks, rather than devoting an entire 1000 warhead second wave to only the 100 remaining targets. This situation puts a premium on producing very reliable missiles and warheads (R near 90%), and on replacing those warheads observed to fail during the attack by firing backup missiles pre-programmed for the missions of the failed warheads. At least for the early failures (during lift-off and boost and MIRV bus phases, i.e. before the individual warheads have separated) it should be feasible to do this by pre-programming an entire set of replacement missiles for the missions of the attacking missiles and firing replacements for only those missiles observed to fail (23,24). The US has apparently followed this tactic at least since 1976 (24). Such a tactic could increase the effective reliability significantly.

For the much more difficult problem of replacement of reentry-stage failures, Richard Garwin has suggested that "bombs that squeak" or other approaches could inform the attacker as to which warheads had failed (29). Pre-programmed missiles (single-warhead perhaps, because it would need to be done on an individual warhead basis) could then replace those warheads that had failed, although the replacements would arrive at least one ICBM flight time, 30 minutes, later.

Such replacement of either early failures or reentry failures, superimposed on a highly effective first wave, might be preferable to a "conventional" two-wave attack, especially in an arms-controlled world.

For the optimistic end, replace the previous 250 m CEP with 125 m, and again assume R=70% and 5% fratricide. This implies SSPS=6%, and an overall 2-shot survival probability Eq.(9) of 14%. Figure 4 shows these results as well one result based on assumptions that are intermediate between the pessimistic and optimistic cases. Table 2 lists all assumptions and results.

Table 2. Survivability of the MM/MX force in the mid-1990s, to a 2-wave (except as noted) attack by Soviet SS-18-type ICBMs. Assumptions and conclusions.

	pessimistic assumptions	intermediate assumptions	optimistic assumptions
SS-18 CEP	90 m	110 m	125 m
SS-18 yield	0.5 MT	0.5 MT	0.5 MT
MM/MX silo strength	130 atm	165 atm	200 atm
SSPS	0.1%	2%	6%
Reliability	90%	80%	70%
Fratricide	0	3%	5%
Fraction surviving	1%	5%	14%
Fraction surv 1st wave	10% (1-wave)		

The future: Attack by SLBMs

Today's Soviet SLBMs are not accurate enough to attack silos. The CEPs of the SS-N-18 and SS-N-20 SLBMs are estimated at 600 m (14,30), although the recently deployed SS-N-23 is thought to be more accurate than this (11,14). Warheads of plausible yield (0.5 MT or less) need CEPs below 300 m in order to destroy hardened silos.

However, the historical 7-year halving time (Figure 5) suggests that mid-90s Soviet SLBM CEPs will be 300 m or less. Furthermore, the US Trident II SLBM planned for deployment in 1989 is expected to have a 120 m CEP, approaching MX accuracy (15,31,32), so the 7-year *lag* time (Figure 5) suggests future Soviet SLBM CEPs as low as 120 m.

Technically, Trident II's accuracy is achieved through an inertial guidance system with one or more position "updates" based on stellar sightings at the end of the boost phase (28,32). Use of the Navstar satellite global positioning system would improve Trident II accuracy even further, but the Navy might prefer not to depend on Navstar because its vulnerability to attack could make it unreliable. Another contributor to US SLBM accuracy is the Navy's Increased Accuracy Program which, in the 1970s, reduced the uncertainties in the initial launch position and velocity, uncertainties that can be quite large for submarine launched missiles (33-35). There is no particular reason why the Soviets can't make similar improvements, attaining something like Trident II CEPs.

In fact, DOD stated recently (13) that "Improved accuracy of the Soviets' latest SLBM systems, as well as possible efforts to increase SLBM reentry vehicle size and warhead yield, would confirm Moscow's plans to develop a hard-target-kill capability for its SLBM force. "

All in all, it is fair to say that the Soviets could plausibly have hard-target-capable SLBMs by the mid-1990s. Such a development could perhaps still be prevented by arms control (36), or it could on the other hand be encouraged by US deployment of Trident II, by US reliance on strategic forces that are especially vulnerable to hard-target-capable SLBMs, and by arms control incentives that favor new SLBM deployments over new ICBM deployments (as US START proposals presently do).

Thus Soviet SLBM CEPs should be down at least to 300 m, and perhaps as small as 120 m, by the mid-90s. Putting these numbers into the optimistic and pessimistic estimates, respectively, and proceeding precisely as in the case of attack by ICBMs, one obtains the results shown in Figure 4 and Table 3. As we in Tables 1 and 2, we also include an intermediate case, midway between the worst case and the best case, in Table 3. Note the wide vulnerability error bars, stemming from the wide uncertainty in future Soviet SLBM CEPs. This wide range of possibilities may tempt the Soviets to work toward SLBM accuracy improvements.

Table 3. Survivability of the MM/MX force in the mid-1990s, to a 2-wave (except as noted) attack by Soviet SLBMs. Assumptions and conclusions.

	pessimistic assumptions	intermediate assumptions	optimistic assumptions
SLBM CEP	120 m	210 m	300 m
SLBM yield	0.5 MT	0.5 MT	0.5 MT
MM/MX silo strength	133 atm	165 atm	200 atm
SSPS	2%	32%	62%
Reliability	90%	80%	70%
Fratricide	0	3%	5%
Fraction surviving	2%	22%	57%
Fraction surv 1st wave	12% (1-wave)		

SLBMs and synergistic survivability

The question of Soviet SLBM accuracy is critical, because of "synergistic survivability" of US ICBMs and bombers. As we have seen, ICBMs could be vulnerable to a Soviet ICBM attack today. The bombers, on the other hand, are vulnerable to a short warning (15 minutes or less) SLBM attack launched from near US coastlines and designed to surprise most bombers at or near their bases. Soviet ICBMs could not attack with such short warning, due to their 30-minute flight time.

Now, suppose that the Soviets attempted to destroy both the ICBMs and the bombers today. If they launched their SLBMs at the same time as their ICBMs, they could destroy most US bombers during the first 15 minutes, but there would be 15 more minutes before the arrival of the first Soviet ICBMs. During this time, US ICBMs could be launched. If the Soviets chose instead to launch their ICBMs 15 minutes prior to their SLBMs so that both would explode simultaneously, US ICBMs could be destroyed in their silos (assuming no launch upon the warning of Soviet ICBM launch, a policy that raises many additional risks and complications) but the bombers would have a full 30 minutes warning between Soviet ICBM launch and the first SLBM explosions at US bomber bases. Briefly, although the ICBM and bomber forces may not be individually survivable today, they are synergistically survivable: Either one or the other force would largely survive and could retaliate. The Scowcroft Commission relied strongly on this synergistic survivability of ICBMs and bombers in its argument favoring immediate deployment of MX missiles in vulnerable Minuteman silos (37).

A Soviet ability to attack ICBM silos with SLBMs will change this situation drastically. If the Soviets had enough accurate SLBMs to attack all US silos, and in addition enough less-accurate SLBMs to attack all US bomber bases, a massive short-warning SLBM attack from near US coastlines could destroy both components. Many observers see this as a significant threat (33,34,38-41). For example, Robert McNamara has pointed out that "our newest submarine force will soon carry missiles accurate enough to destroy Soviet missile silos. When the Soviets follow suit, as they surely will, their offshore submarines will, for the first time, pose a simultaneous threat to our command centers, our bomber bases and our Minuteman ICBMs" (38). Feiveson and Duffield argue that SLBMs fired close to US shores are well suited to attacks on US strategic bomber bases, control facilities, and ICBM silos, and that the reassurance provided by synergistic survivability "will evaporate once the Soviets develop the capability to strike both missile silos and bomber bases simultaneously with SLBMs" (41).

Thus the Soviets may have the capability, and vulnerable US ICBMs may give them the incentive, to develop a silo-destroying SLBM.

Some may question the ability of the Soviets to mount the massive off-shore submarine deployment that would be needed for a short warning attack on both ICBMs and bombers. Sixteen submarines would be needed to attack the 1000 US ICBM silos alone, assuming double-targeting. It can be argued, for example, that such a congregation of submarines near US shores would be detected, and US forces alerted and anti-submarine warfare forces deployed. However, if the Soviets chose to pursue an SLBM-attack policy based on hard-target-capable SLBMs, they would begin to station many submarines off US coasts routinely. As Feiveson and Duffield state, "none of these problems [of keeping large numbers of Soviet ballistic missile submarines on station] are clearly insurmountable (41)." Continuing vulnerability of US ICBMs gives the Soviets some incentive to pursue this strategy.

Strategic considerations: deterrence, pre-emption, stability, uncertainty

It must be emphasized that this chapter studies vulnerability only of the ICBMs, while deterrence is based on the ability of the full triad of US strategic forces to retaliate for an attack. ICBM vulnerability does not equal strategic vulnerability. Even in the worst case considered above, the case of a successful Soviet SLBM attack on both US ICBMs and bombers, the submerged submarines (somewhat more than half of US submarines under normal peacetime conditions, and a truly formidable deterrent force) would still survive. The survivability of any significant portion of even one leg of the triad would surely prevent any rational Soviet leader from launching a first strike except in the most dire of circumstances.

But it is precisely "dire circumstances," the crisis situation, that strategic stability is all about. Essentially all observers, this one included, agree that an unprovoked Soviet "Pearl Harbor" surprise attack (or, in the Soviet view, a "22 June 1941" attack by the US) is out of the question, and is the *least* plausible way that a nuclear war could begin (see, e.g. (42,43)). It is the crisis situation, the escalation of a lower-level crisis into a superpower confrontation, that is thought to be the *most* plausible scenario.

Soviet military thinking appears to hold that, in a crisis, the US may threaten to use nuclear weapons in order to pressure the Soviet Union, that the US might actually carry out that threat with a first strike against Soviet forces, and that therefore the Soviets must be prepared to pre-empt in order to get in the all-important first blow in any future superpower confrontation (20,44-54). For example, "Soviet history and domestic politics prevents any Soviet leader from advocating a military doctrine whose guiding principle is accepting the enemy's first blow--a repeat of the disaster of 22 June 1941. --The Soviet perspective can be summarized as: never again by surprise, never again on our territory" (48). The Soviets appear to have genuinely feared a US strategic first strike during the Cuban missile crisis (46).

The Soviet Union's goals in any such preemptive first strike in a crisis are probably the obvious ones: to limit damage to themselves in the coming unavoidable catastrophe, and to conclude the exchange with some semblance of "victory," or at any rate to avoid defeat.

The distinction between a true surprise attack and preemption in a crisis is crucial. The Soviets are keenly aware that, in any strategic nuclear exchange, terrible damage to themselves would be unavoidable, and in fact this is precisely why no serious observers believe they would launch an unprovoked surprise attack. But in a crisis in which the US has for example begun preparing nuclear weapons for launch, the Soviets may believe that nuclear war is highly likely regardless of whether they attack first. In this case, the question of whether there will be a massive US retaliation becomes less important than the question of whether the inevitable US attack will be significantly reduced, and the chances of Soviet "victory" significantly enhanced, by Soviet preemption. If, for example, the Soviets believed that their civilian casualties might be reduced from over 100 million to "only" a few tens of millions, this could provide the incentive for a damage-limiting first strike.

How might the Soviets reduce damage to themselves? Clearly, by attacking all the vulnerable US forces that are capable of attacking them, especially the strategic forces. In fact, the Soviets could even hope to reduce damage to relatively "small" levels (a few million dead, for example) by destroying nearly all of the vulnerable US strategic systems (bombers and ICBMs and submarines in port) plus the command and control system, and hoping that command and control difficulties in the aftermath of nuclear attack would then prevent most retaliation from the remaining forces and especially from the submarines. I emphasize that this could only be a hope, that they certainly could not count on such a weakened retaliation. But, facing a probable (in Soviet eyes) nuclear attack from the US, this could be their best hope.

The point of maintaining stable, invulnerable nuclear forces that can neither threaten Soviet strategic forces nor themselves be threatened is precisely to prevent either side from having such a damage-limiting option.

Our analysis emphasizes the uncertainties, the "vulnerability error bars," in the outcome of an attack. The usual view is that such uncertainties deter a potential attacker, because a strategic attack would surely not be launched unless it was essentially certain that essentially zero nuclear retaliation would occur (3,4,6,55,56). For example, it may be argued that the 9-49% survivability of today's MM/MX force would by itself deter the Soviets, even assuming that the Soviets were willing to accept retaliation by 90 ICBMs (9% of the present force), because as much as 49% of the MM/MX force *might* survive, and this large fraction would surely devastate the Soviet Union.

But this argument misses the crucial distinction between surprise attack and preemption in a crisis. In an already-threatening situation in which there was a serious threat of nuclear war, or of some other Soviet disaster such as loss of control in Eastern Europe, a preemptive strike might seem to the Soviets to be the best move if there was even a plausible *chance* (not 100% certainty) that damage could be reduced *significantly* (not necessarily to near zero). That is, a wide vulnerability error bar deters surprise attack, but may do little to deter preemption in a crisis, especially one in which the Soviets felt threatened.

If we accept this view of strategic uncertainties, then Figure 4 shows that US ICBMs are today unstable against preemptive attack, because in a crisis the Soviets *might* destroy as many as 910 of them leaving "only" some 90 for retaliation. Depending on the fate of US bombers and SLBMs, this could imply significant Soviet damage reduction, although obviously not to nearly zero. In a sufficiently threatening crisis, the Soviets could find it "least bad" to take the risk and try to destroy those 910 US ICBMs in order to reduce their expected damage.

In any case, if we accept the more general view that crisis instability is one of the more plausible causes of nuclear war, then Figure 4 shows that US ICBMs will be unstable against Soviet ICBM attack, and perhaps against Soviet SLBM attack as well, in the mid-1990s.

Summary and conclusions

Under pessimistic "worst case" assumptions, the MM/MX force is 91% vulnerable to a 2-wave ICBM attack today, and will in the mid-1990s be essentially 100% vulnerable to either ICBM or SLBM attack, perhaps even single-wave attacks, using only a small fraction of Soviet warheads. This situation could bode ill for future crisis stability, even if there are significant uncertainties in these predictions, for in a threatening situation the Soviets might reason that a preemptive attack should be risked if there is even a plausible *chance* of destroying essentially all US ICBMs. On the other hand, the possible survival of the bomber leg, and the probable survival of the submarine leg of the US triad, might discourage any preemptive attack in a crisis.

Furthermore, MM/MX vulnerability to SLBM attack could cause the Soviets to try for preemptive destruction of not only US ICBMs, but also the bombers, in a massive SLBM attack. If we take such a pessimistic forecast seriously, we must contemplate the possibility that the US would be left with only one independently survivable leg of the triad, namely the submarines.

On the other hand, equally plausible but more optimistic assumptions imply that the MM/MX force is only 51% vulnerable to ICBM attack today, and will be "only" 86% vulnerable to ICBM attack and only 43% vulnerable to SLBM attack in the mid-1990s.

Briefly, US ICBMs may be highly vulnerable to ICBM attack today. In the future, they will be highly vulnerable to such attack, and may in addition be highly vulnerable to SLBM attack.

On the basis of these conclusions, many would argue that the US must deploy new land-based strategic forces. The purpose of this study is to investigate the most-mentioned alternatives.

On the other hand, many would argue that there is no need to deploy new strategic forces, that a diad of subs and bombers is plenty, with or without 1000 increasingly vulnerable silo-based Minutemen and MXs. The diad option will be studied in Chapter 5.

References and notes

1. Kosta Tsipis, *Arsenal*, Simon & Schuster, New York (1983), Appendix D.

2. Kosta Tsipis, "Nuclear explosion effects on missile silos," Report for the Center for International Studies, MIT, Cambridge, MA, Feb 1978.

3. Matthew Bunn and Kosta Tsipis, "Ballistic missile guidance and technical uncertainties of countersilo attacks," Report No.9, Program in Science for International Security, MIT, Cambridge, MA, Aug 1983.

4. Matthew Bunn and Kosta Tsipis, "Uncertainties of pre-emptive attack," Scientific American, Nov 83, 33-41.

5. C.S. Draper, "Imaginary problems of ICBM accuracy," letter to New York Times, 20 Sep 1981.

6. J. Edward Anderson, "First strike: myth or reality," Bulletin of the Atomic Scientists, Nov 1981, p. 6-11.

7. Samuel Glasstone and Philip J. Dolan, *The Effects of Nuclear Weapons*, Third Edition, US Department of Defense, Washington, D.C. (1977).

8. Anibal A. Tinajero and Louis C. Finch, "Cost to attack US and Soviet strategic forces under three alternative arms control approaches," Congressional Research Service Report, Library of Congress, 23 May 1985.

9. George Keyworth, quoted in Jane's Defense Weekly, 29 September 1984.

10. Les Aspin, *Midgetman: Sliding Shut the Window of Vulnerability*, House Armed Services Committee, Washington, D.C., 10 Feb 1986.

11. *Soviet Military Power* . Department of Defense, Washington, D.C , (1984, 1985).

12. ibid., 1987.

13. ibid., 1988.

14. Barton Wright, *Soviet Missiles: Data From 100 Unclassified Sources*, Lexington Books, Lexington (1986).

15. Thomas B. Cochran, William M. Arkin, Milton M. Hoenig, *Nuclear Weapons Databook, Vol. I: US, Nuclear Forces and Capabilities*, Ballinger Pub. Co., Cambridge, MA (1984).

16. "Soviet silos can survive nuclear strike," Defense Daily, 15 April 1985, p. 249.

17. Walter B.Slocombe, "Strategic Forces, " in *American Defense Annual 1985-1986*, edited by G. Hudson and J. Kruzel, Lexington Books, Lexington, Mass. (1985), pp 77-96.

18. Jane Boulden, "Who's ahead? Examining the nuclear balance," Background Paper No. 12, Canadian Institute for International Peace and Security, Mar 1987

19. Congressional Budget Office, "Trident II missile test program," Staff Working Paper, Feb 1986.

20. Robert Bermanand John Baker, *Soviet Strategic Forces: Requirements and Responses* The Brookings Institute, Washington, DC (1982).

21. Tsipis-3 - Kosta Tsipis, "The operational characteristics of ballistic missiles," SIPRI Yearbook 1984, Taylor & Francis, London (1984).

22. Bruce Bennett, "How to assess the survivability of US ICBMs," Rand Report R-2577-FF, June 1980.

23. Ashton Carter, "Satellites and anti-satellites: the limits of the possible," International Security, Spring 1986, 46-98.

24. John D. Steinbruner and Thomas M. Garwin, "Strategic vulnerability; the balance between prudence and paranoia," International Security, Summer 1976. 138-181.

25. "Targeting flexibility emphasized by SAC," Aviation Week and Space Technology, 10 May 1976, pp. 29-34.

26. Dietrich Schroeer, *Science, Technology and the Arms Race*, John Wiley & Sons, Inc., New York (1984).

27. Jonathan Medallia, "MX, Midgetman, Minuteman and Titan missile programs, Congressional Research Service Issue Brief, 26 Mar 1987.

28. Robert Alderidge, "Trident-I and first strike," Ground Zero, Winter 1987.

29. Richard Garwin, letter to *Physics and Society*, Oct 1988.

30. International Institute of Strategic Studies, *The Military Balance 1984-1985*, IISS, London.

31. "Strategic force upgrade," Aviation Week and Space Technology, 9 March 1987, p. 37.

32. Peter Clausen, Allan Krass, Robert Zirkle, *In Search of Stability: an assessment of new US nuclear forces*, Union of Concerned Scientists, Cambridge, MA (1986).

33. Joel Wit, "American SLBM: counterforce options and strategic implications," Survival, July 1982, pp. 163-174.

34. Robert S. Norris, "Counterforce at sea: The Trident II missile," Arms Control Today, Sept 1985, pp. 5-10.

35. William M. Arkin, "Sleight of hand with Trident II," Bulletin of the Atomic Scientists, Dec. 1984, p. 5-6.

36. Michael Dukakis, quoted in Arms Control Today, Jan/Feb 1988, p. 9.

37. *Report of the President's Commission on Strategic Forces*, Chaired by Brent Scowcroft, April 1983, Reprinted by the Library of Congress.

38. Robert McNamara, quoted in The Defense Monitor, Vol. 16, Number 6 (1987), p. 3.

39. Karl Lautenschlager, "The submarine in naval warfare, 1901-2001," International Security, Winter 1986-87, p. 94-140.

40. US Senate Foreign Relations Committee, Hearing on 11 May 1983: Brent Scowcroft's testimony states "--for now and perhaps for a decade [i.e. until 1993] in the future we have what I would call a synergistic survivability between the bomber and the ICBM forces,--."

41. Harold Feiveson and John Duffield, "Stopping the sea-based counterforce threat," International Security, Summer 1984, p. 186-202.

42. NEXT Magazine, "A survey into the outlook for nuclear war 1980 - 1990, Summer 1980 (Survey of 25 experts.)

43. Graham Allison, Albert Carnesale, Joseph Nye, Jr., editors, *Hawks, Doves, & Owls,* W. W. Norton & Co., New York, 1985

44. Robert McNamara, *Blundering into Disaster*, Pantheon Books, New York (1986).

45. Franklyn Griffiths, "New thinking in the Kremlin," Bulletin of the Atomic Scientists, April 1987, p. 20-24.

46. Marc Trachtenberg, "The influence of nuclear weapons in the Cuban missile crisis," International Security, Summer 1985, p. 137-163.

47. Daniel Ford, *The Button*, Simon and Schuster, New York, 1985.

48. Stephen Meyer, "Soviet strategic programmes and the US SDI," Survival, Nov/Dec 1985, p. 274-292.

49. Dan Strode, "The Soviet Union and modernization of the US ICBM force," Chapter 8 of *Missiles for the Nineties*, ed. by B. Schneider, C. Gray, and K. Payne, Westview Press, Boulder and London, 1984.

50. Stephen Shenfield, "The Soviet undertaking not to use nuclear weapons first and its significance," Detente, Oct 1984, p. 10-11.

51. John Erickson, "The Soviet view of deterrence: A general survey," Survival, Nov/Dec 1982, p. 242-251.

52. Willliam M. Arkin, "The drift toward first strike," Bulletin of the Atomic Scientists, Jan. 1985, p. 5-6.

53. Raymond Garthoff, "The Soviet SS-20 decision," Survival, Mar/April 1983, p. 110-119.

54. Michael MccGwire, *Military Objectives in Soviet Foreign Policy*, Brookings Institution, 1987.

55. Stanley Sienkiewicz, "Observations on the effect of uncertainty in strategic analysis, " World Politics, Oct 1979, 90-110.

56. Steve Smith, "MX and the vulnerability of American missiles", ADIU Report, May/June 1982, pp. 1-5.

Part V Chapter 4

Hard point defense of land-based missiles*

Ruth H. Howes

Introduction

Strategic defensive systems fall into two broad categories. Area defenses are systems that protect all targets within a relatively large area of a nation. For example, an area defense might protect all targets in the four states of Montana, Wyoming, North Dakota and South Dakota. Clearly it is not practical to station individual interceptors near every target in a large area where there are many military and economic targets. Therefore area defenses are targeted on all missiles heading toward a large area of the country to intercept them soon after launch or while they are still hundreds of kilometers from their targets. Consequently they provide a measure of protection not only for hardened military targets in the region but also for soft civilian targets such as factories and population centers. Any defense based in space to attack missiles during the boost phase of their flight will have some capability as an area defense as do ground-based long range exoatmospheric interceptors such as those proposed for the Accidental Launch Protection System.

Hard point defenses are designed to protect specific hardened military targets such as a single flight of missile silos or a communications center. They are usually stationed on the ground near the facilities they are designed to protect and have limited ranges. Beam weapons are limited to targets in their line of sight while rocket-powered interceptors are limited by the acceleration they can achieve in order to intercept an incoming warhead after it has been identified. Such defenses can protect hardened facilities but they offer little protection to cities or economic assets in the region of the military targets because of their limited range. In an extreme case, one might design defenses to protect a flight of ten missile silos and offer no protection to the crew in the launch control center nearby. A typical deployment of hard point defenses would be to protect the MX Missiles deployed in Minuteman silos near F. E. Warren Air Force Base in Wyoming.

Ground-based interceptors designed to protect either a group of missile silos or a single city have been extensively studied by both the United States and the Soviet Union. Both superpowers have deployed defensive systems utilizing ground-based interceptors although only the Soviet Galosh system is currently operational (1). The problems in deploying such defenses are better understood and analyzed than those involved in deploying space-based defenses and consequently it is possible to make more realistic estimates of probable effectiveness and costs (2). Certainly ground-based systems are technically less complex than untried space-based systems since deployed systems give the analyst a basic structure for which to study effectiveness.

The ability of either the United States or the Soviet Union to deploy defenses in space that are effective in attacking missiles during the boost phase of their flight is a matter of considerable debate. The current consensus of the technical community is that a "perfect" defense capable of intercepting all Soviet warheads targeted on the United States and western Europe is not feasible at the present time. According to the Office of Technology Assessment, "Assured survival of the US population appears impossible to achieve if the Soviets are determined to deny it to us" (3). The goal of space-based defenses being studied by the Strategic Defense Initiative Office in the late 1980s is to provide limited protection for military and economic assets. Even former Secretary of Defense Caspar Weinberger, an early supporter of a completely effective defensive shield, was speaking in 1987 of a "phased introduction of each element of a whole systems" (4).

The shape which such limited defenses should take is a matter of open controversy. The 1987 *Report of the American Physical Society Study Group on Directed Energy Weapons* summarizes the technical issues of using beam weapons for area defenses. This lengthy report deals only with laser and particle beams. It does not touch on other types of defensive systems or on the strategic implications of deploying defenses, but it clearly demonstrates the complexity of the technical issues in designing area defense systems. It equally clearly exposes our inability to answer these technical questions at the present time (5). The 1988 Office of Technology Assessment report *SDI: Technology, Survivability and Software* underlines similar concerns for SDI software and battle management systems. It also highlights the vulnerability of defensive systems to potential Soviet countermeasures and attacks on the defenses themselves (6).

Strategically, point defenses are stabilizing in the sense that any system that ensures the survivability of the land-based leg of the Triad further deters the Soviets from considering a first nuclear attack. Since point defenses have very limited ability to protect large areas such as extended population distributions against nuclear attack, their only effect is to increase the survivability of land-based missiles and very limited population concentrations such as the center of Washington or a single factory in the face of a Soviet first strike. This increases Soviet hesitation in launching such an attack just as Soviet point defenses would cause us to hesitate in attacking them first. If they are deployed around missile silos, point defenses are strategically analogous to super-hardened silos or mobile missiles since they tend to stabilize the nuclear arms race against a nuclear first strike.

From a strategic point of view, area defenses are inherently very different. In a world of partially effective area defenses, the side which attacks first with nuclear weapons suffers less damage than the side which waits to respond. Thus the effect of deploying defenses with an area capability is likely to be to place the nuclear arms race on a hair trigger. Any effort to deploy defenses may fuel an offense-defense spiral in the arms race. Furthermore, deployment of defenses at more than one site clearly violates any interpretation of the Anti Ballistic Missile (ABM) Treaty.

Because they are technically feasible and well-studied and because their strategic effect is analogous to that of other technologies studied here, terminal defenses will be treated as meaning point defenses designed to protect missiles based in silos. No space-based defenses will be considered here since any space-based system has at least a limited effectiveness as an area defense. Point defenses will be subdivided into traditional ABM defenses such as those deployed by both superpowers and based on ground-launched rocket-powered interceptors and novel defenses which are schemes designed largely to protect only very hard targets such as missiles in silos. Novel defenses are still speculative and have not yet been deployed. They are infrequently discussed in the open literature. Traditional ABM defenses have some capability as area defenses since they involve exoatmospheric interceptors with considerable range but their primary strategic goal is to protect a very limited set of targets.

Traditional ABM defenses

Systems for defending hardened targets consist of apparatus performing four essential functions. The first function is to track the enemy warhead throughout its flight. Defenses must warn of impending attack and realize that incoming nuclear warheads are approaching. They must discriminate between attacking warheads and decoys or natural phenomena and track reentry vehicles to determine their trajectories. After the defense has attacked the warhead, the defenses must be sure that the warhead has indeed been destroyed so that there is no need to take another shot at it.

The second function of defenses is to launch a weapon at the warhead and guide it to its target. Obviously sensors that can track a reentry vehicle have some capability to guide an interceptor to it. Alternatively interceptors can be equipped with on-board sensors which track the incoming warhead.

The third mission of a defensive system is to disable or destroy its target. A final mission is to control the entire system of defenses so that two interceptors aren't fired at one incoming warhead while another is not targeted at all. Interceptors will have to obtain information from tracking systems and the tracking data will have to be channelled to the interceptor that is going to attack that warhead. Sensors will have to decide whether or not to intercept targets in the first place and which targets are decoys and which are warheads. All these activities, which are collectively known as battle management, will probably be conducted at least partially by complex computer systems.

Soviet ballistic missile defense

The only defense against ballistic missiles deployed today is the Soviet Union's Galosh System which protects Moscow. The Soviets have deployed defenses against ballistic missiles since the beginning of the technology (7). They have recently completed a massive upgrade of the Galosh system adding more sophisticated interceptors and battle management radars. In its present form, Galosh will receive initial warning of US launch of its ICBMs from a satellite detection system giving defenders about thirty minutes warning time to prepare for an attack and rough data on the origin of the missile (8). The Soviets have deployed two over-the -horizon radars which could provide as much as thirty minutes warning of the arrival of intercontinental ballistic missiles. A new over-the-horizon radar is under construction to scan the Pacific Ocean (9).

The real workhorses of the Soviet system for detecting incoming nuclear missiles are 11 Hen House radars with ranges of 6000 km deployed at 6 locations on the periphery of the Soviet Union (10). The original Hen House network is currently being supplemented by the addition of six new large phased array radars. Five of them are deployed on the periphery of the Soviet Union and will improve the coverage already provided by the Hen House network. The sixth, which has been constructed at Krasnoyarsk in Siberia, closes a large gap in the Soviet early warning system. It is 750 kilometers from the nearest border and is oriented across approximately 4000 kilometers of Soviet territory to the northeast. Thus if it is an early warning radar designed to detect and track ICBMs, it is a clear violation of the ABM Treaty, which requires warning radars to be located on the borders of nations, oriented outward. The Soviets claim it is designed for tracking space objects, but since its design is identical to other Soviet large phased array radars designated as early warning radars such as that at Pechora, and since it conveniently closes the last gap in the Soviet early warning system, most observers feel it is designed for warning of a nuclear attack.

Large phased array radars (LPARs) are electronically rather than mechanically steered. Hence they can distinguish and track multiple targets simultaneously. The exact capability of the Soviet system cannot be known without an examination of the computers that steer the beam of the LPAR and process the data it receives, but the new early warning system now being deployed will certainly have the ability to confirm data from the satellite network and the over-the-horizon radars and to gather tracking data on incoming missiles which could be passed to smaller radars which would guide interceptors to destroy the attackers. LPARs are massive objects which take years to construct. They are generally considered an essential element of any defense against ballistic missiles. Clearly they have capability to support area defenses for a major portion of the Soviet Union as well as point defenses for Moscow. It should also be noted that LPARs are vulnerable to nuclear attack since they are necessarily soft targets and are not easy to repair quickly. They operate at VHF frequencies and are subject to black out from the effects of nuclear explosions (7).

As incoming warheads approach Moscow, they will be picked up by Dog House (range 2,800 km (10)) and Cat House battle management radars located south of Moscow. These older phased-array radars have recently been supplemented by a new four-sided phased-array radar known as Pill Box at Pushkino north of the city. All of these phased-array radars have ranges of hundreds of kilometers characteristic of the long range target acquisition radars. The original GALOSH system included four complexes each of which contained two installations. Each installation contained a large target-tracking radar, the TRY ADD radars, which were large, mechanically steered dishes which probably operated on targets acquired by the LPARs. The installations also contained two smaller mechanically steered dishes probably used for guiding interceptors and eight huge nuclear-armed GALOSH interceptors. The purpose of the interceptor was to attack its target well above the atmosphere at extremely long range (7).

The new Moscow ABM system consists of seven new launch complexes which will contain two types of interceptors. The modified GALOSH missile will still be designed for interception at ranges greater than 320 kilometers above the atmosphere (10) and is believed to nuclear armed. The GAZELLE is a new high acceleration missile probably designed to intercept its targets within the atmosphere at ranges on the order of tens of kilometers. The system is expected to have the full complement of 100 interceptors that can be legally deployed under the ABM Treaty. The old TRY ADD mechanically steered radars will be replaced by the Flat Twin modular phased array radar for target tracking and the Pawn Shop phased array radar for missile guidance (11). Because Pawn Shop is mounted in a van shaped container, it could presumably be moved in a matter of weeks if its container had wheels. Similarly Flat Twin modules could be disassembled and moved in months.

The Reagan administration has argued that this possible mobility of the phased array radars constitutes a "potential violation" of the ABM Treaty (12) although no one claims to have seen the Soviets actually move them once they have been deployed. It is not clear that Pawn Shop and Flat Twin will actually be deployed since they seem to have been removed recently from Soviet test ranges leading some observers to conclude that they were part of an unsuccessful development program (13). In this case, the GALOSH upgrade will operate with radar technology similar to that of the US Sentinal/Safeguard system which was abandoned as inefficient.

A second treaty compliance question which arises with respect to the GALOSH upgrade is whether or not the launchers in the system are capable of being rapidly reloaded with spare missiles since the Soviets have been observed to reload test silos at the Sary Shagan Test Range in less than a day. The US

Government currently feels that the situation should be of concern because rapid reloading of launchers would greatly increase the capability of the GALOSH system but finds it "ambiguous" as regards its legality under the ABM Treaty (12).

In addition to the BMD network around Moscow, the Soviet Union has traditionally deployed strong air defenses around its territory. The defenses consist of fighter interceptors, extensive radars and surface to air missiles (SAMs). In particular, the newest Soviet SAM, the SA-12 is probably designed to shoot down aircraft, cruise missiles and tactical ballistic missiles. Its long range, high altitude version, the SA-X-12B, which is currently being tested may have some capability against strategic ballistic missiles (9).

Levi and O'Hanlon have conducted an extensive study of the capability of modern SAMs for defense against ballistic missiles. Their analysis shows that if one makes the most generous assumptions about the performance of these air defense systems, one can calculate that they might individually have some ability against ballistic missiles but their performance could be easily degraded by decoys and other countermeasures or overwhelmed by large numbers of warheads. The Soviet air defense would be more capable in defending hardened silos but would still have minimal impact in protecting them against a US attack at current levels of arsenals (14).

US defenses against ballistic missiles

The United States currently has no operational defenses against ballistic missiles. However a defensive system, the Sentinal/Safeguard system, was developed and deployed around missile silos at Grand Forks, North Dakota in the early seventies. In 1975, the system was mothballed because it was very expensive to operate and of extremely limited utility as a defense. The Sentinal system is a two-layer system very similar in design and capabilities to the updated GALOSH system. It utilized large phased array radars known as Perimeter Acquisition Radars to track incoming missiles with initial acquisition taking place approximately 4000 kilometers or 10 minutes down range of the target. If the missile were on a lofted or depressed trajectory, warning times would be shorter.

The first layer of the defenses, the Spartan interceptors would be launched to intercept the incoming warheads six hundred kilometers down range. The Spartan was a three stage, solid-fueled rocket which carried a nuclear warhead designed to detonate well above the atmosphere. The Spartans were to be based on "farms" alerted by the network of acquisition radars deployed along the northern border of the US. The radars handed targeting information to tracking radars located near the farms and used to guide the Spartans to their targets. The system was unable to distinguish decoys from RVs until they re-entered the atmosphere and limited in the number of objects it could track in real time.

The second layer of the Sentinal system was the short range, high acceleration Sprint missile. Sprint was a two-stage solid fuel missile designed to intercept warheads at altitudes below 35 kilometers. It carried a nuclear warhead in the kiloton range as compared with the megaton warhead of the

Spartan. Sprint farms were to be located next to Spartan farms and equipped with Missile Site Radars whose function was to guide both the Spartans and the Sprints to their targets. The Missile Site Radars and the Perimeter Acquisition Radars were soft targets whose functioning was key to the success of the defense. The major purpose of Sprint was to protect the radars and the Spartan interceptors from an attack directly on them (15). Neither Sprint nor Spartan carried any guidance or tracking system on board the interceptor itself.

The Achilles' heel of the BMD systems proposed in the early seventies and based on conventional interceptors like Sprint and Spartan is clearly the guidance systems for leading the interceptors to their targets. Although they are faster to steer and far more capable than the original mechanically steered radar dish-shaped antennas, electronically steered phased-array radars are very expensive. Currently deployed systems are immobile and soft targets for nuclear weapons. Because of their cost, they cannot be proliferated and they cannot be hidden since target acquisition radars must be operated constantly to detect a surprise attack. Tracking radars must operate during a nuclear battle if defenses are to be effective so they must be relatively insensitive to nuclear effects..

In addition to nuclear detonations, large phased array radars are vulnerable to attack by conventional weapons guided by radar-seeking systems such as the US's HARM missile. They can be jammed, deceived by decoys and blacked out for extended periods of time by such effects of distant nuclear blasts as the expanding ionization of the fireball or the electromagnetic pulse from the burst. Exact mechanisms for the blackout depend on the relative location of the radar and the explosion, the altitude of the detonation and the wavelength at which the radar operates. Once a defense like Sentinal has lost its radar eyes, it cannot fight blind since its interceptors cannot find specific targets but must shoot at random.

The tracking capacity of the radars of the early seventies was severely limited by the computer capability needed to process radar data in real time. Modern computers have significantly improved data processing capability and thus greater ability to track multiple objects in real time and to run discrimination algorithms.

A final drawback to the widespread deployment of the Sentinal system was the dislike of the American public for nuclear-armed interceptor missiles based within their communities necessary for defense of cities.

Current technological trends in conventional defense of missile silos

Both the United States and the Soviet Union have large research programs aimed at upgrading their defenses against ballistic missiles. In the U.S., the program was originally lodged in the Army's anti-tactical ballistic missile program and since 1983 has been incorporated in the Strategic Defense Initiative (SDI). Although SDI seeks to develop defenses which will attack incoming nuclear missiles in all phases of their flight, many of its most successful technologies can support a purely hard point defense. The current extensive discussion of immediate deployment of SDI focuses on using defenses of limited effectiveness to protect the land-based leg of the Triad, but it should be emphasized that the proposals of such groups as the Marshall Institute (16) focus on space-based systems which will attack missiles during the boost phase of their flight.

In the area of target acquisition and tracking, the Soviet Union is upgrading the system of radars which scan beyond its borders as discussed in connection with the GALOSH system. The United States and Canada have mounted a joint program known as the North American Air Defense Modernization Project which provides for construction of a North Warning System to replace the thirty year old DEW (Distant Early Warning) radar line in Canada. Phase one will consist of 13 radar installations with 200 mile range which are scheduled for completion at the end of 1988 and will stretch close to the 70th parallel from Alaska to Labrador. Phase two plans to fill in gaps between the long range radars of phase one with 39 radars of range 70 miles. The solid state radars are designed to operate unmanned and relay maintenance information to a central location. Although the short range radars are probably designed to track aircraft and cruise missiles approaching from the arctic, they will also have some capability to track ICBMs on polar trajectories. The radars will be linked via satellite to NORAD, the main command center for North American Air Defense under Cheyenne Mountain in Colorado.

New over-the-horizon backscatter radars are being built along the northern coasts of the US and will cover approach routes south of the arctic ionization which blocks propagation of over-the-horizon radar. AWACS, airborne radar planes, will be based in Canada and will improve coverage of the region in the event of an alert (17).

The Canadians are working on developing a space-based radar for improved tracking and coverage of attacking ballistic missiles, cruise missiles and airplanes (18), while the US is developing a ground-based tracking and targeting radar, the Terminal Imaging Radar (19). The Airborne Optical Adjunct being developed by SDI will be an infrared telescope mounted in a Boeing 767 capable of tracking incoming reentry vehicles. Because it is mobile, it is more survivable in the face of nuclear attack than ground-based radars. According to SDI, the AOA is not a component of an ABM system because it is a test of the feasibility of infrared tracking and discrimination and not a mobile component of an ABM system which is forbidden by the ABM Treaty (19). The development of sophisticated computers for battle management, which is one of the major programs of SDI, will undoubtedly produce more capable battle management systems for hard point defense. The Soviets are also believed to be pursuing advanced computer technology.

To improve the ability of interceptors to destroy incoming warheads, two major areas have been investigated: non-nuclear interceptors with on-board guidance systems to track their targets, and beam weapons.

The High Endoatmospheric Defense Interceptor (HEDI) is a guided, high-velocity missile which will destroy its targets well inside the atmosphere using some form of conventional blow such as a spray of pellets or a chemical explosive (19). The Exoatmospheric Reentry Vehicle Interception System (ERIS) is also guided to its target which it destroys by conventional means above the atmosphere. ERIS type vehicles have been flight tested, for example in the army's Homing Overlay Experiment of 1984 (20). The guidance systems considered include infrared seekers, optical trackers and radar tracking systems. The Army's FLAGE (flexible lightweight agile guided experiment) (21) test demonstrated the ability of a radar guided missile to intercept a tactical ballistic missile. It should be pointed out that all such tests of experimental systems are conducted by highly trained crews under ideal weather conditions and that failures of preliminary tests are not reported. The FLAGE program had been succeeded by the Extended Range Intercept Technology program which will use a FLAGE type interceptor modified to receive midcourse guidance from ground based radars (22).

Both the United States and the Soviet Union have major research programs on the use of lasers and particle beams to intercept missiles. The US program is currently part of SDI and includes a test site at White Sands, New Mexico. The primary goal of laser and particle beam research within SDI has been to develop weapons to attack missiles in the boost phase. The technology has been reviewed by a number of good sources including Ashton Carter, The Office of Technology Assessment, and the American Physical Society Study Group on Directed Energy Weapons (23).

Laser and particle beam weapons can also be effective in attacking reentry vehicles in the terminal phase of their flights. The problem of transmitting a beam through the atmosphere is being considered in connection with the use of ground-based lasers whose beams would bounce off space-based mirrors to kill their targets. The Soviet research program on beam weapons is centered on their test range at Sary Shagan. According to the Department of Defense (24), the test range contains several laser weapons designed for air defense and two larger test systems, at least one of which might have some capability against reentry vehicles. The Soviets may also have other sites for testing laser weapons. Photographs taken by the French commercial satellite SPOT have identified a construction site apparently to be used for military lasers 30 miles southeast of Dushanbe, the capital of the Tadzhik Republic, in the southern part of the USSR (25).

Progress in developing particle beam weapons is frequently mentioned but the research effort is not as clearly evident as the one on laser weapons. At the present writing, none of these laser weapons has been tested against an ICBM. Each side can state believably that its beam weapons are designed to attack satellites. There is no treaty forbidding research into anti-satellite weapons, so these claims place laser research well within the range of activities allowed by existing treaties. To state the obvious, the fact that a defensive weapon is being tested by no means indicates that it is ready for deployment. For example, candidates for the Army line of sight-forward-heavy air defense system scored only six hits in fifteen attempts to shoot down helicopter drones which move much more slowly than incoming reentry vehicles (26).

Novel terminal defenses

The final category of silo defenses consists of relatively simple defensive systems which are designed to protect only missiles in hardened silos and offer almost no defense to other targets in the vicinity such as command and control stations or radars. They have no area defense capability and are strategically comparable to hardening silos.

All the conventional systems described above are designed to track incoming missiles at great distance from the protected target and to attack them either above the atmosphere or in the high reaches of the atmosphere in order to minimize damage to ground based targets. For example, hardened radars can withstand an overpressure of 10 to 100 psi. This means that they might be vulnerable to one megaton detonations at optimal height of burst within a ground range 10 kilometers. A soft radar would require protection from much lower overpressures since radar dishes are easily damaged and can be blown over (27). These figures reflect only the effects of blast damage on the systems. Thus traditional terminal defenses seek to protect not only the missiles in their silos but the basic command and control systems which operate them. In the case of defensive batteries, this becomes critical since the defensive interceptors are guided to their targets by the radars that are themselves vulnerable to enemy attack.

Novel defenses assume that the object of the defense is to protect only the missiles in their silos. Thus they can operate with a very small keep-out volume.

A typical novel defense for silos is a dust defense in which nuclear explosives are detonated to kick up a cloud of dust and small gravel in the path of an incoming RV. Since the RV reenters the atmosphere at nearly 7 km/s, it has a high probability of being destroyed by collision with the debris between it and its target(28). The dust defense allows incoming missiles to approach within a few kilometers of their targets before they are destroyed so that ground targets are vulnerable to incoming warheads fused to detonate on attack. If nuclear explosives are used to produce the dust clouds, and in most schemes for this type of defense warheads on the order of 1 megaton are needed to produce a large enough cloud, the defense of US missile fields will involve detonation of tens of megatons of nuclear warheads on US territory although a well-designed defense will produce minimum fallout, blast and thermal damage to its surroundings. The cost of a false alarm in a dust defense is high to the defender who will needlessly detonate many megatons on his own territory.

Other novel defenses include ideas such as pellet swarms launched in the path of incoming missiles or the use of clusters of machine guns to attack RVs at very short distances from defended targets. Such defenses leave soft targets near the defended target vulnerable to nuclear explosions from warheads which are designed to detonate on attack. Novel defenses will probably need to be fired by remote controls since protection of crews near the target site poses a problem for these defenses.

A second series of novel defenses use the effects of nuclear detonations to destroy incoming RVs. The best publicized of these is the closely spaced basing scheme proposed for the MX missile in which missiles would be based so close together that detonation of a nuclear warhead in an attack on one silo would destroy an incoming RV targeted on its neighbor. Other proposals have included detonation of a "friendly" warhead above a silo field in order to destroy incoming enemy warheads by the dust and debris kicked up or by the neutron flux from the initial detonation. These defenses not only produce nuclear detonations on friendly territory but it is not certain that they will work over a protracted attack.

They might well be frustrated if the silos in the field could not withstand a close nuclear detonation or if the enemy timed his attack to shoot between defensive detonations.

A final defense of missiles in silos would be to launch them on warning. This option is described and evaluated in Part IV Chapter 2 of this study.

Conclusions

Partially effective point defenses of missiles based in silos can be constructed using existing technology. Present day technology has improved the ABM systems of the 1970s in three major areas. First, radars can be proliferated and some might be made mobile so the tracking systems for defensive interceptors are no longer sitting ducks for enemy attack (29). The reason for these improvements lies in the advances made in solid state electronics so that both individual radar elements and the computer systems that control them can be made cheaper and lighter. Second, the use of self-guiding interceptors has increased the capability of the interceptor to get close to its target so that the defense may no longer need to rely on nuclear detonations to destroy incoming warheads.

Third, the rapid advance of computer technology has made strides in developing schemes for managing a nuclear battle. Unlike area defenses for population centers, hard point defenses for land-based missiles do not have to work perfectly but merely must preserve a sufficient number of land-based missiles to convince the enemy that the use of nuclear weapons against the United States would leave the US able to launch a devastating counterattack against the attacker's hardened military targets and his civilian command centers. Application of these technical advances to BMD components is still in the early stages of development, and it should be noted that the technology is anything but mature.

Smart interceptors, large mobile radars and sophisticated computers for battle management are expensive, and careful estimates of the costs of specific defensive systems will have to be balanced against the cost of comparable protection obtained using mobile missiles or superhardened silos. Detailed cost estimates will depend on the numbers of interceptors to be deployed which in turn is driven by estimates of the number of enemy RVs that will be used in an attack on the protected missiles. A frugal defender can expand the effectiveness of his defenses if he can base interceptors to defend only selected silos and conceal his choice of defended silos from the enemy. The enemy would then have to attack all silos as if they were defended and pay a price in attacking warheads equal to the attack cost if all silos were defended. A further cost of deploying defenses would be the destruction of the ABM Treaty.

In the face of a determined enemy, any defense can ultimately be overwhelmed by a superior offense. Proliferation of attacking warheads and decoys can exhaust defensive interceptors. Even beam weapons can attack a limited number of targets at the same time so a wealthy and determined enemy can simply saturate the defense. Of course the defense can continually increase the number of interceptors it deploys. In a situation without treaty constraints where both sides are willing to spend almost unlimited sums on weapons, the effectiveness of defenses will depend on their cost effectiveness at the margin, that is how much it costs to deploy the last interceptor relative to the cost of deploying yet one more offensive warhead.

In addition to proliferating offensive weapons, the offense can work to deploy countermeasures designed to frustrate the defense. Typical countermeasures might include lofting or depressing the trajectories of ballistic missiles to reduce the warning time given to the defense. Extra ablative coatings can cover RVs to force beam weapons to remain longer on a single target. Decoys can complicate tracking and battle management functions. Radars can be jammed, and the heat of high altitude nuclear detonations used to blind infrared sensors. Finally the defense must expect the offense to attack it directly so that defenses like the Sentinal/Safeguard system which rely on soft, expensive stationary sensors like large phased array radars, are ineffective in a real nuclear attack. In assessing the value of any defensive scheme, the defender must imagine the countermeasures that the offensive will employ against it and be aware that he may not have identified all of them.

With the current thaw in Soviet-American relations, it seems likely that the world of the future will involve arsenals that are restricted by treaties. In this case, defenses seem more likely to be effective than in a world of unlimited arsenals since the offensive cannot simply saturate the defense. This is also the case for mobile basing modes since limitations on nuclear weapons restrict the area that can be subjected to a barrage attack. In the event of reductions to a very small number of strategic warheads, say perhaps 1000 on each side, hard point defenses might play a critical role in insuring the survival of a substantial fraction of these warheads and thus be effective in deterring nuclear attack. Marginal changes in the effectiveness of defenses might prove critical in improving the guarantee of nuclear retaliation following a first strike. Thus it seems likely that future arms control regimes would include restrictions on deploying hard point defenses although the present limits of 100 interceptors based at a single site may be relaxed. Testing of hard point defenses at allowed test ranges is permitted by the ABM Treaty. Certainly a workable agreement on large cuts in offensive warheads would include limitations on deploying and developing area defenses.

A second future development that seems virtually certain is that the accuracy of offensive missiles will improve. Special purpose warheads seem certain to be deployed in the future. One such special purpose warhead might be an earth penetrator which digs into the ground near a silo before exploding in order to crush the silo more efficiently. These advances in technology seem likely to threaten any silo basing mode no matter how hard the silos are made. Point defenses could counteract these advances in technology if the number of offensive warheads that could be deployed were limited by treaty.

Mobile basing modes are not vulnerable to increasing accuracy because the effectiveness of a barrage attack depends only on the total equivalent megatonnage that is delivered to the region. They are vulnerable to detectors which can track mobile systems in near real time so that they can be targeted as they move. The flight time from the Soviet launch sites to the US is approximately half an hour, but near real time tracking data would greatly reduce the area that would have to be barraged, particularly for heavy, slow-moving missile transporters. In comparing the value of hard point defenses and mobility in ensuring survivability of land-based missiles, the probability of such advances in tracking technology must be considered.

The advantages of deploying terminal defenses in order to ensure the survival of a substantial portion of the land-based leg of the triad will thus depend on the treaty constraints assumed for the world of the future, the predictions of the course of technological advance, and the relative political and economic costs of deploying defenses and offenses.

References and notes

* This work was partially supported by a Faculty Summer Research Grant from Ball State University.

1. *Jane's Weapon Systems 1985-86,* Jane's Publishing Company, London (1985).

2. US Office of Technology Assessment, *Ballistic Missile Defense Technologies,* US Government Printing Office, Washington, D.C.(1985), pp. 198-201.

3. ibid., p. 33.

4. Casper Weinberger, "Why Offense Needs Defense," Foreign Policy, Volume 68 (1987), p. 16.

5. Study Group on Science and Technology of Directed Energy Weapons, *Report to the American Physical Society*, Reviews of Modern Physics, Supplement to Volume 59 (1987).

6. US Office of Technology Assessment, *SDI: Technology Survivability, and Software*, US Government Printing Office, Washington, D.C. (1988).

7. Sayre Stevens, "The Soviet BMD Program," in *Ballistic Missile Defense*, edited by Ashton B. Carter and David N. Schwartz, The Brookings Institution, Washington, D.C. (1984) pp. 182-220.

8. Data on the Soviet BMD system around Moscow is taken from *Soviet Military Power 1987*, US Government Printing Office, Washington, D.C. (1987) pp. 45-53 and from *Soviet Strategic Defense Programs*, Department of Defense and Department of State, Washington, D.C. (1985) pp. 7-13.

9. *Soviet Military Power 1987*, p. 60.

10. The International Institute for Strategic Studies, *The Military Balance 1984-1985* , IISS, London (1984).

11. National Campaign to Save the ABM Treaty, *The Threat of Soviet Breakout from the Anti-Ballistic Missile (ABM) Treaty*, National Campaign to Save the ABM Treaty, Washington, D.C. (1985).

12. *Soviet Noncompliance*, US Arms Control and Disarmament Agency, Washington, D.C. (1986).

13. Arms Control Association , *Soviet Strategic Defense Programs*, Arms Control Association, Washington, D.C. (1987).

14. Barbara G. Levi, and M. O'Hanlon, "Assessing the Capabilities of Soviet Air Defenses for Ballistic Missile Defense," preprint.

15. Richard L. Garwin and Hans A. Bethe, "Anti-Ballistic-Missile Systems," Scientific American, March, 1968; reprinted in *Arms Control*, introduced by Herbert F. York, W.H. Freeman and Company, San Francisco (1973), pp.164-174. Yield data are taken from George W. Rathjens, "The Dynamics of the Arms Race," Scientific American, April, 1969. reprinted in the same work, pp. 177-187.

16. John Gardner, Edward Gerry, Robert Jastrow, William Nierenberg, and Frederick Seitz, *Missile Defense in the 1990s*, George C. Marshall Institute, Washington, D.C. (1987).

17. "North Warning Radar on Schedule but Phase 2 Faces Delays," Aviation Week and Space Technology, 28 Sep 1987, pp. 132-133.

18. "Canada Regards Space-Based Radar as Follow-On to North Warning System," Aviation Week and Space Technology, 28 Sep 1987, pp. 136-137.

19. Ivo H. Daalder, "A Tactical Defense Initiative for Western Europe," Bulletin of the Atomic Scientists, May 1987, pp. 34-39.

20. Ashton B. Carter, "The Relationship of ASAT and BMD Systems" in *Weapons in Space*, edited by Franklin A. Long, Donald Hafner and Jeffrey Boutwell, W. W. Norton and Company, New York (1987), p. 174-175.

21. John D. Morrocco, "Army Missile Test Demonstrates FLAGE Guidance," Aviation Week and Space Technology 1 Jun 1987, pp. 22-23.

22. US Office of Technology Assessment (1988), p. 110.

23. Ashton B. Carter, *Directed Energy Missile Defense in Space-A Background Paper*, US Office of Technology Assessment, Washington, D.C. (1984). The other two reports are cited in notes 1 and 3 above.

24. *Soviet Military Power 1987*, pp. 50-51.

25. William J.Broad, "Private Satellite Photos Offer Clues About Soviet Laser Site," New York Times ,23 Oct 1987

26. John D. Morrocco, "Army Air Defense Candidates Score Six Hits in 15 Attempts," Aviation Week and Space Technology, 21 Sep 1987, pp. 24-25.

27. Stephen Weiner, "Systems and Technology," in *Ballistic Missile Defense*, edited by Ashton B. Carter and David N. Schwartz, The Brookings Institution, Washington, D.C. (1984), pp.49-97.

28. John Michener, "First Order Design of Large Dust Sources for Defense of Hard Sites Against Ballistic Missile Attack," preprint; also Weiner, p. 89-91.

29. US Office of Technology Assessment (1988), p. 96.

Part V Chapter 5

The diad alternative

Art Hobson

Introduction

For over two decades, both superpowers have possessed a "triad" of strategic forces: land-based ICBMs, sea-based SLBMs, and bombers carrying gravity bombs, short-range attack missiles (SRAMs), and air-launched cruise missiles (ALCMs) (see Tables 1 and 2 of Chapter 1).

This three-fold multiplicity, although partly an accident of history, serves several purposes. The Scowcroft Commission, which in 1983 conducted a broad high-level overview of US strategic modernization, emphasized three purposes (1). First and foremost, each leg acts as a hedge against failure of the other two legs by having its own independent deterrent effect. Ideally, each leg should be sufficiently survivable and powerful to ensure deterrence by itself. Thus, in order to prevent a devastating US retaliatory strike, the Soviets would have to solve three separate problems. It is unlikely that Soviet breakthroughs could be simultaneously this successful.

The second purpose is to "force the Soviets, if they were to contemplate an all-out attack, to make choices which would lead them to reduce significantly their effectiveness against one component in order to attack another" (1). The leading example of Soviet attack trade-offs is the "synergistic survivability" of US ICBMs and bombers, discussed in Chapter 3: although the ICBM force and bomber force may be individually vulnerable, timing problems preclude a simultaneous attack on both.

Third, the three legs complement each other militarily because each component has military advantages not present in the others. Table 1 lists the military pros and cons generally given for the triad components. The evaluation in Table 1 is sanctioned by traditional military thinking which holds, for example, that it is always good to be able to destroy the enemy's weapons. But such traditional thinking might be inappropriate for strategic nuclear weapons, because the foremost purpose of these weapons is deterrence, not winning. This question is discussed further below.

As seen in Chapter 3, US ICBMs might be vulnerable today, but present trends in Soviet missiles would make them certainly vulnerable in the mid-1990s. Thus, if the US is to maintain a viable triad through the remainder of the century, it must decide today what to do about ICBM vulnerability.

But do we need a triad? Would an air and sea *diad* provide equal or greater security? After all, the number "three" is not inscribed in stone. One truly survivable component would be enough. For example, there was little threat of Soviet pre-emption during the 1950s, despite the fact that the US had no ICBMs or SLBMs at all. In this chapter, I study the diad option.

It is striking that the diad alternative is so seldom considered. Although ICBM vulnerability began to be discussed in the mid-1970s, and the missiles began to be vulnerable in the early 1980s, and despite the failure on the drawing boards or in Congress of one land-basing scheme after the other, few in the government or in the defense department ask "do we really need ICBMs?". The reason may be that the vulnerability problem is always posed in the form "how can we preserve a survivable land-based missile force?" rather than "how can we preserve US security in light of land-based missile vulnerability?". Thus the Scowcroft Commission, whose deliberations initiated the present MX and Midgetman programs, does not mention the diad alternative in its final report. This neglect is probably explained by the way in which the President originally posed the problem: he "asked us (the Commission) to examine the future of our ICBM forces and to recommend [land] basing alternatives" (1).

It seems obvious that, if land-based missiles are becoming obsolete, one question that needs to be asked is: Do we need them?

A notable exception to the diad "allergy" is the Office of Technology Assessment's *MX Missile Basing Study* (2) which includes, among 11 plausible options for future ICBM basing, moving off the land onto small submarines, surface ships, or airplanes. In fact two of these three options, small subs and airplanes, are among OTA's five leading options (the other two are multiple protective shelters, ABM defense of multiple protective shelters, and launch under attack). For others who explicitly question the need to maintain land-based ICBMs, see (3-8). For strong arguments on the other side, see (9,10). For presentations of both points of view, see (11-13).

Most triad/diad arguments are related to the three purposes of the triad stated above. I discuss each of these purposes in turn, focusing on the ICBM component.

Table 1. Traditional military pros and cons of triad components.

	PROS	CONS
ICBM:	Highly accurate, destroys silos Good command and control Fast, 30 minutes	May be vulnerable today Vulnerable by mid-1990s Not recallable after launch
SLBM:	Excellent endurance, months Invulnerable when submerged Good deterrent and reserve force Will soon be highly accurate Very fast, 15 min or less	Poor command and control Communication slow, tenuous, dangerous Not recallable after launch
BOMBER:	Good command and control Recallable after take-off ALCMs and SRAMs highly accurate Flexible targeting	Slow, several hours Medium endurance, days Vulnerable to short warning attacks Vulnerable to air defenses

Independent survivability

Do the ICBMs actually have an independent deterrent effect, as a hedge against failure of the other legs? Do we need a hedge, i.e. might the other legs fail?

The answer to the first question is "probably not." ICBMs have an *independent* (as constrasted with synergistic, below) deterrent effect only if they are independently survivable. Independent survivability of the ICBMs is precisely what is vanishing, and what this study is all about.

Can ICBMs then be made survivable? The verdict, seen throughout this study, is: Not necessarily. Certain new ICBM basing modes will be moderately survivable, independently, *if and only if* coupled with appropriate arms control measures.

Vulnerable ICBMs, far from acting as a hedge against failure, might actually make a negative contribution to deterrence by encouraging pre-emptive Soviet attack during a crisis. As we argued in Chapter 3, a pre-emptive attack against any vulnerable nuclear weapons, to reduce damage in an already-threatening situation, is the most plausible scenario for strategic nuclear war.

Concerning failure of the other legs: The non-alert two-thirds of the bomber force has always been vulnerable to a short-warning attack. Even the remaining alert portion might have limited survivability under SLBM attack, especially if depressed trajectories were used. In a crisis, the entire bomber force would be put on alert and reaction time would be reduced, *if* there were sufficient warning and *if* the President did not hesitate to give the alert command. But in the worst case (short warning), the bombers are independently vulnerable (14).

US submarines are widely acknowledged to be invulnerable at least throughout this century (15-18). If one could be absolutely certain of this prediction, and if there were no other reasons to doubt the ability of submarines to retaliate for a first strike, then deterrence could be preserved by the submarines alone. As far as deterrence is concerned, we could dispense with both ICBMs and bombers. But of course we cannot be certain of this prediction. If a Soviet breakthrough in anti-submarine warfare did occur, the US would certainly want another, less vulnerable, strategic force to fall back on. Furthermore, there is another critical problem beyond submarine survivability: The order to retaliate might not get through because communication with submerged submarines is slow and tenuous, although US submarines could decide to fire on their own if they sensed that a first strike had occurred.

Some feel that the US needs at least one other, independently survivable, hedge against failure of the SLBM leg. In light of bomber vulnerability, a survivable ICBM force would be highly desirable. But the problem is that survivable ICBMs might be impossible. Given enough weapons, or the right kinds of weapons, the Soviets can make any land-basing scheme vulnerable.

In the absence of 50% cuts and a ban on particularly threatening weapons such as maneuvering reentry vehicles (MaRVs, Chapter 11), enhancements of a bomber/submarine diad would provide a more realistic hedge against Soviet technological breakthroughs than would new additions to the ICBM force. In fact, the addition of ALCMs to the bomber force and of SLCMs to the sea-based (submarine and surface) force are enchancements of precisely this sort. Some analysts (4,19) regard the ALCMs and SLCMs as essentially another, fourth, leg of US strategic forces.

The OTA study of MX basing, discussed above, analyzed three such enhancements of a bomber/submarine diad: long-range missiles on surface ships, in small submarines, and on aircraft. Unfortunately, the consideration of such options has seldom gotten to the point of serious research and development.

Synergistic survivability

Is synergistic survivability of ICBMs and bombers important? Will it be maintained?

Until about 1980, the US had two invulnerable triad legs and one more-or-less vulnerable leg: the bombers, which have always been fairly vulnerable to a sufficiently determined Soviet attack. During the early 1980s, the ICBMs began to be vulnerable also. Today the US has one invulnerable leg and two vulnerable legs.

Synergistic survivability mitigates any real or peceived dangers arising from the individual vulnerability of two triad legs. Although accurate Soviet ICBMs can destroy the US ICBM force, and short-warning SLBMs can destroy the bomber force, timing problems make it impossible to execute the two attacks together. Either US ICBMs or bombers would get away. In effect, the US has a "diad" today: an invulnerable ICBM/bomber leg on the one hand, and an invulnerable SLBM leg on the other. Each of these two legs is independently survivable, and each acts as a hedge against the failure of the other. The existence of these *two* independently survivable components is plenty, in my opinion, to preserve deterrence.

The problem is that the ICBM/bomber leg depends on synergistic survivability, and this will vanish if the Soviets deploy an equivalent to the US Trident II SLBM (20). With a Trident-II equivalent, the Soviets would be able to attack US ICBM silos with high-accuracy SLBMs while attacking bombers with lower-accuracy SLBMs. It was suggested many times during the past few years that arms control restrictions on the development of high-accuracy SLBMs be instituted to prevent this problem (21-24). The advice was not taken. It may be too late now. The Trident II is in the middle of its flight test program, and funds are being appropriated for fiscal year 1989 for procurement of operational missiles. Initial deployment is planned for December 1989. Trident II will be able to attack Soviet silos, and it can only be a matter of time before the Soviets deploy a similar missile. The Department of Defense states that new Soviet SLBMs "are likely to be more accurate and possess greater throw-weight than their predecessors, and they may eventually provide the Soviets with hard-target-capable SLBMs" (25). From the point of view of survivability, US forces would then effectively form a diad with one leg vulnerable: a vulnerable ICBM/bomber force on the one hand, and the submarines.

Can new land-based missiles fix this problem? We will see in Chapter 11 that superhard-silo-based Midgetmen would solve it, assuming that the Soviets do not develop the demanding capability needed to destroy *superhard* silos with SLBMs.

But mobile missiles might not solve the problem. SLBMs may be the most plausible way to attack mobile missiles. If new US ICBMs are mobile, then ICBMs and bombers will be synergistically survivable only if the required SLBM attack is so large as to be implausible or impossible. This in turn depends on the size of the Soviet SLBM arsenal. The Department of Defense expects this force to grow from 20% of today's Soviet strategic warheads to over 30% of an expanded force by 1995 (25). As we will see in Chapter 7, such a force is large enough to attack the entire US ICBM force (silo-based MM/MXs, and mobile missiles) along with the bombers, provided it is done with short warning. Cuts of 50% or more in the Soviet arsenal would prevent this situation, especially if the new mobile missile was random-mobile Midgetman rather than dash-mobile Midgetman or rail-based MX.

It should be added that all of this assumes that the Soviets won't develop MaRVs (Chapter 11), especially MaRVs that can be launched by submarines. MaRVs could make a mobile missile force, or a superhard silo force, vulnerable to a smaller number (i.e. much smaller than would be needed for an area barrage) of attacking SLBMs or ICBMs.

Military capabilities

Does the US need the military capabilities offered by ICBMs (Table 1)? To what extent are these capabilities duplicated by the other legs?

To answer the second question first, all three of the listed ICBM capabilities are possessed by the other legs, but they are not all possessed by any one other leg. By the mid-90s or sooner, the US SLBM force will be able to destroy essentially all 1400 Soviet silos in a 1-wave attack with 15 minutes warning, i.e. submarines will be able to carry out the ICBM's silo-destruction mission. But submarines do not offer fast or secure command, control and communications (we call these, collectively, "control"). The bombers, on the other hand, offer good control, and their ALCMs are even more lethal than the MX (26), but they require several hours to reach their targets.

Briefly, all three components can destroy silos: SLBMs can do it quickly, bombers can do it with good control, but only ICBMs can do it quickly *and* with good control.

But does the US need this "prompt controlable silo-destruction capability"? A deterrent posture implies that the US would only launch a strategic strike second, in retaliation for a Soviet first strike, in which case Soviet silos would be empty. The Scowcroft Commission (1) and others reply that a prompt silo-destruction capability is needed in order to be able to fight (i.e. in order to deter via the threat of being able to fight) a limited nuclear war in which the Soviets hold many of their ICBMs in reserve. This reply is not convincing: any Soviet ICBMs held in reserve in a limited nuclear war would surely be on launch-on-warning status and would be launched prior to the arrival of US warheads.

It is hard to think of a military reason for a prompt silo-destruction capability, if the purpose of US strategic weapons is deterrence.

In fact, a prompt silo-destruction capability actually *detracts* from US security by pushing the Soviets to preempt in a crisis out of fear of US preemption. Prompt silo-destruction capability is de-stabilizing, no matter which superpower possesses it. To deter nuclear war, what is needed is survivability on both sides, not silo-destruction capability.

Other arguments

Many other arguments have been made for retaining the ICBMs, in addition to the three triad arguments given above.

Supporters of a prompt and controllable silo-destruction capability have argued (1,9,10,27,28) that "In order to deter such Soviet threats [as for instance a limited use of Soviet nuclear weapons against other countries] we must be able to put at risk those types of Soviet targets--missile silos, nuclear weapons and the rest--which the Soviet leaders have given every indication by their actions they value most, and which constitute their tools of control and power" (1).

In this view, the Soviets' most valued assets are their missile silos, in contrast to the United States' most value assets, which are its people (28). To my knowledge, nobody has ever offered any evidence for this depressing view of the value structure of the Soviet leadership. Soviet military preparations and actions do not support this view any more than parallel US military preparations and actions support a view that our highest values reside in our missile silos. And even if this view were correct, it does not answer the military fact that a US silo-destruction capability is useless because in a crisis Soviet missiles are likely to be on launch-on-warning status, and our ICBM warheads would find only empty silos.

It is argued that the US needs a prompt and controllable silo-destruction capability because the Soviets have one: "Effective deterrence of any Soviet temptation to threaten or launch a massive conventional or a limited nuclear war thus requires us to have a comparable ability to destroy Soviet military targets, hardened and otherwise. --A one-sided strategic condition in which the Soviet Union could effectively destroy the whole range of strategic targets in the United States, but we could not effectively destroy a similar range of targets in the Soviet Union, would be extremely unstable over the long run" (1).

But the ability of Soviet missiles to destroy US silos has never been a one-sided affair: The US Minuteman IIIA has a lethality equal to that of the SS-18 mod 4, and the two missiles were deployed at about the same time. And although there are only 900 MM 3A warheads as opposed to 3000 SS-18 warheads, the MM 3As can attack 4000 of the highly MIRVed Soviet warheads while the SS-18s can attack only 2000 US warheads. Thus from the point of view of the targets, the US has always had a 2-to-1 edge. Furthermore, if there ever was a Soviet-dominated asymmetry, it has been at least partly rectified by the US ALCM and will be further rectified by Trident II, so US ICBMs may not be necessary for this purpose.

More fundamentally, new silo-destruction capability is always destabilizing, regardless of whether the other side already has this capability, for the reasons already discussed.

One school of thought holds that basing a triad leg specifically on US soil is essential for deterrence. It is said that "the US would convey an image of weakness by letting the Soviets drive US ICBMs from the land" (11), i.e. land-based missiles have symbolic value. A more rational argument is, with no US ICBMs, the Soviets would be able to launch a preemptive strike against US strategic forces without having to attack targets in the continental United States, so they might be less deterred than they have been in the past. On the other hand, US bomber and submarine bases are also on US soil, and they would have to be attacked in a preemptive strategic strike.

Discussion and conclusions

In sum, the good reasons for preserving a land-based ICBM component are, in my opinion: independent survivability of ICBMs, synergistic survivability of bombers and ICBMs, and concerns over loss of command and control over US SLBMs. The prompt and controllable silo destruction capability possessed by ICBMs such as the MX, far from being an advantage, actually makes the US less secure by pushing the Soviets to launch a damage-limiting pre-emptive strike in a crisis.

But high-accuracy Soviet ICBMs have foreclosed the independent survivability of ICBMs. And high-accuracy US SLBMs make it only a matter of time before the Soviets get the same capability and synergistic survivability vanishes also.

As seen throughout this study, there is no basing scheme that will allow the United States, acting alone, to guarantee either the independent or synergistic survival of its land-based missiles. Land-based missile survival can be ensured only with the cooperation of the Soviets, through arms control. Total missile arsenals must be reduced by 50% or more, and threatening developments such as MaRVs and earth-penetrators must be banned, or there are no land-based missile schemes that make any sense. Vulnerable strategic missiles are of little help, and indeed they probabably make a negative contribution to deterrence.

In the absence of arms control to ensure land-based missile survival, US security would in my opinion be best-served by foregoing new ICBM deployments and relying instead on a diad of submarines and bombers. For example, the warheads now on the vulnerable MM/MX force would provide far more security if they were deployed on either of the other two legs.

It is not the purpose of this study to investigate the many plausible ways by which a bomber/submarine diad could be enhanced. Certainly the ALCM and SLCM deployments already made or planned are enhancements of this sort. Other possibilities have been put forward in the past. They deserve serious consideration.

References and notes

1. *Report of the President's Commission on Strategic Forces,* chaired by Brent Scowcroft, April 1983, reprinted by the Library of Congress.

2. Office of Technology Assessment, *MX Missile Basing,* Congress of the U.S., Washington, D.C. 1981).

3. Gerald E. Marsh, "Danger of limited SDI," Bulletin of theAtomic Scientists, March 1987, pp. 13-14.

4. Stansfield Turner "The folly of the MX missile," The New York Times Magazine, 13 Mar 1983, pp. 84-85, 94-96.

5. Senator Sam Nunn, quoted in Arms Control Today, April 1986, p. 17.

6. William Perry, interview in "William Perry and the weapons gamble," Science, 13 Feb 1981, pp. 681-683.

7. Boston Study Group, *The Price of Defense*, Times Books, New York, 1979, pp. 78-79.

8. David C. Morrison, "ICBM vulnerabaility," Bulletin of the Atomic Scientists, Nov 1984, pp. 22-29.

9. Colin Gray, "Why does the U.S. need ICBMs?" NATO's Fifteen Nations, Aug/Sept 1982, pp. 80-84.

10. Blair Stewart, "The Scowcroft commission and the 'window of coercion'," Strategic Review, Summer 1983, pp. 212-27.

11. Jonathan E. Medalia, "Assessing the options for perserving ICBM survivability," Congressional Research Service Report, 28 Sept 1981.

12. A.A. Tinajero, "Analytical framework for strategic force otpions: the MX ICBM and its alternatives," Congressional Research Service Report, 1 April 1983.

13. Stephen J. Cimbala, "Do we need a land-based missile force?" National Defense, Oct 1985, 46-52.

14. Edwar N. Luttwak, *Strategic power: military Capabilities an dpolitical utility*, SAGE Policy Paper No. 38, Center for Strategic and International Studies, Georgetown University (1976), p. 66.

15. Norman Polmar, " That other leg in the triad," Air Force Magazine, July 1985, pp. 84-91.

16. Richard L. Garwin, "Will strategic submarines be vulnerable?" International Security, Fall 1983, pp. 52-67.

17. "US-Soviet nuclear arms 1985," Defense Monitor, Vol. 14, No. 6 (1985).

18. John Lehman, Sec. of the Navy, quoted in Union of Concerned Scientists Fact Sheet, 1983.

19. Carnesdale and Charles Glaser, "ICBM vulnerability: the cures are worse than the disease," International Security, Summer 1982, pp. 70-85.

20. *Soviet Military Power 1988*, U.S. Department of Defense, Washington, D.C. (1988).

21. Harold Feiveson and John Duffield, "Stopping the sea-based counterfore threat," International Security, Summer 1984, pp. 187-202.

22. Robert S. Norris, "Counterforce at sea: the Trident II missile," Arms Control Today, Sept 1985, pp. 5-10.

23. Donald MacKenzie, "Missile accuracy - an arms control opportunity," Bulletin of the Atomic Scientists, June/July 1986, p.11-16.

24. "First strike weapons at sea," Defense Monitor, Vol. 16, No.6 (1987).

25. *Soviet Military Power 1987*, U.S. Department of Defense, Washington, D.C. (1987).

26. Thomas B. Cochran, William M. Arkin, Milton M. Hoenig, *Nuclear Weapons Databook, Vol. I: US Nuclear Forces and Capabilities,* Ballinger Pub. Co, Cambridge, MA (1985).

27. Interview with Keith B. Payne, "On nuclear strategy: the MX, strategic nuclear talks, and limited nuclear war," Global Perspectives, Dec 1983.

28. Barry Schneider, Colin Gray, Keith Payne, editors, *Missiles for the 90s,* Westview Press, 1984.

Part V Chapter 6

Command and control of land missile basing modes*

Ruth H. Howes

Introduction

The consideration of new superhard or mobile basing modes for land-based missiles is prompted by predictions that current silos can be destroyed by newly deployed, accurate Soviet missiles. In order to be useful for a retaliatory attack on the Soviet Union, missiles must survive a Soviet attack with their propulsion and guidance systems in working order. In addition to the missile itself, the system must have survivable command and control which will allow launch of the missile against an enemy target.

This essay examines command and control issues for each of the basing modes described in this study. We study the requirements for command and control systems in light of the missile's strategic mission. Also we describe the current and planned command and control systems for US land-based missiles.

Requirements for command and control of land-based missiles

The strategic function of a weapons system determines its command, control and communications (C3) requirements. Since the strategic mission of the various components of the land based leg of the U.S. nuclear triad is the subject of ongoing debate, the precise requirements for its command and control are difficult to determine.

The small intercontinental ballistic missile (Midgetman) was formally proposed by the President's Commission on Strategic Forces (1) in its report of April, 1983 as one leg of a three part proposal which included basing the large MIRVed MX or Peacekeeper missile in Minuteman silos and an effort to limit total warheads on each side by arms control agreements. According to Weinberger's Report to the Congress for Fiscal Year 1987, the Midgetman "promises to complicate Soviet attack planning and enhance survivability in an era of increasingly accurate Soviet missile forces"(2). The United States' declared policy allows neither a preemptive nuclear strike nor launch on warning for strategic systems although the US has refused to pledge no first use of nuclear weapons to stop a Soviet conventional attack in Europe. Therefore the stated function of the land based missile system implies that it is designed to survive a nuclear attack by the Soviet Union and respond to that attack in a timely manner. The Scowcroft Commission also made C3 issues its highest priority for strategic modernization stating: "Our first defense priority should be to ensure that there is continuing, constitutionally legitimate, and full control of our strategic forces under conditions of stress or actual attack "(3).

Positive control of a missile allows the missile to be launched at the discretion of military commanders toward the target selected by these commanders. A minimal operational requirement is that a substantial portion of the land based missile force can be fired following a Soviet nuclear strike. The strategic mission assigned to a missile determines the requirements for its command and control systems.

Many current doctrines require the missile to have the capability to destroy hardened military targets in a protracted nuclear exchange where escalation is controlled. Thus the missile is designed with hard target kill capability. If it is to be used in an attack on hardened and/or time-urgent Soviet military targets, its C3 system must allow for rapid retargeting and flexible, accurate, time of launch so that land based missiles can conduct retaliatory nuclear strikes in co-ordination with other surviving elements of the US strategic nuclear forces. Such retaliatory strikes require control of the missile launch by the National Command Authority (NCA) following a Soviet nuclear strike. The NCA must receive updated information on the status of Soviet forces and relay targeting information to surviving missiles. The missiles would then be retargeted and launched in a coordinated attack on Soviet forces. Of course the NCA would also need updated information on the readiness of US forces. The MX missile has the requisite accuracy to attack hardened targets and thus seems to be designed for use in such war fighting strategies.

Other strategies see the role of land based missiles as riding out a Soviet first strike and attacking targets of value to the enemy in a retaliatory strike designed to destroy the Soviet society. In this case, the attack can be preplanned and only a launch command need be communicated to the missile launch control centers. A launch order could even be transmitted before most of the attacking warheads had reached their targets. For either a prompt launch under attack response or a delayed response, the NCA will need no reports on damage to the US and Soviet nuclear forces and will not need to communicate targeting information and launch timing data to missiles.

The requirements for communications links are far less demanding than those of the warfighting scenarios. Since the guidance system of the Midgetman consumes a substantial portion of its throw-weight, the accuracy of this new missile is an issue in the debate over how to base it. A more accurate guidance system allowing attacks on hardened targets will increase the weight of the missile and influence practical basing modes as well as increasing the demands on communications and control systems. Cochran and his co-authors estimate that the weight of the Midgetman will vary between 25,000 pounds and 35,000 pounds depending on the guidance system used with the payload (4).

Negative control of a missile consists of safeguards to prevent its accidental or unauthorized launch. For example, the permissive action links (PALs) used to protect nuclear weapons in Europe or the elaborate procedures and multikey arrangements used to prevent unauthorized launch of submarine-launched ballistic missiles are negative controls. For mobile basing modes, negative control will include added procedures to protect the missile from accident or terrorist activities.

An obvious tension exists between negative and positive launch controls. Tradeoffs between the two exist for every basing mode. For instance, the importance of negative control of the Pershing II missile based in a mobile mode in Europe requires that the nuclear warheads for the missiles be stored separately from the missiles themselves. This basing scheme prevents accidental launch of nuclear-armed missiles or destruction of the warheads in an accident with possible resultant contamination of friendly territory.

During an accident on January 11, 1985, the solid fueled motor of a Pershing II missile burned killing three GIs(5). The missile carried no nuclear warhead. Had the warhead been mounted on the missile at the time of the accident, the plutonium and uranium it contained might have been distributed over the countryside by the explosion of the rocket fuel or of the explosives contained in the warhead.

However, separate basing of warheads and missiles slows the response time for the Pershing since the warheads would have to be mounted on the missiles before they could strike at Soviet targets. In a surprise attack, this might be a difficult procedure to effect.

Basing modes and C3 issues

Several variants of land mobile basing have been suggested for the Midgetman (6) including those that were discussed at length during the MX basing debate (7). The C3 systems of US silo-based missiles are currently vulnerable to nuclear attack and, at the present time, it seems unlikely that the missiles could be controlled by the NCA during a nuclear exchange. The strategic modernization now in progress will take steps to remedy some of the vulnerability of the C3 systems. The Appendix to this chapter describes details of current C3 systems for missiles based in silos and the C3 component of the Strategic Modernization Program.

C3 for missiles based in superhardened silos

Missiles are gaining high enough accuracies to land within 100 meters of silos. Superhardened silos are designed to survive an attack by a nuclear weapon just outside the cratering radius with the missile in condition to be launched. Presumably a basing scheme using superhardened silos will also use superhardened launch control centers (LCCs) buried deeply near the missile. Cables connecting the silo and the LCC will be difficult to harden within the cratering radius but might be protected by very deep burial. Hardening communication links between the silo and the NCA poses several severe problems. A hardened, deeply buried system of cables between the NCA and the LCCs for the missiles would be expensive, particularly since it would have to be redundant in order not to be cut at some point and very large in geographic coverage. Thus missiles based in superhardened silos will probably have to communicate with the NCA using some portion of the electromagnetic spectrum.

Antennas must be exposed in order to function and are very soft targets for nuclear weapons since they are easy to blow over. Thus antennas will probably have to be deployed after the initial nuclear strike. One model would bury the antenna which could then dig itself out following a nuclear attack. After a nuclear attack on a missile field, the atmosphere in the vicinity may be disrupted sufficiently so that surviving antennas will be unable to send or receive messages for several hours. In these circumstances, NCA authorities will have no way of knowing whether or not the crew of the LCC has survived the attack. Thus superhardened silo basing will probably include provisions for remote launch of the missiles by airborne or land-mobile command posts. Land-mobile command posts are subject to barrage attacks and their vulnerability follows the discussion of the vulnerability of mobile launchers given elsewhere in this study.

If command links cannot be hardened to survive an attack that does not destroy the missile in its superhard silo, the missile cannot be used in a war-fighting scenario but must carry out preplanned retaliatory strikes. If the launch codes are released to the crew of the missile's LCC on warning of a nuclear attack or just at the beginning of an attack, the crew could carry out a preset launch sequence but this scenario places heavy demands on warning systems since a false warning that triggered the launch of missiles would be disastrous. Because warning times can be no more than 30 minutes, release of the launch codes might have to be carried out automatically by a computer system at the time of warning.

Alternatively, the President would have to authorize their release to individual LCC crews at some point during a deepening international crisis and send a brief coded message to trigger the release of the missiles. Nervous, low-ranking officers would thus be physically able to launch ICBMs stationed in superhard silos. Either automated release of launch codes (under attack or on warning of an attack) or early release to LCC crews during an international crisis increases the possibility of the accidental initiation of war. Therefore if superhard silos are used to stabilize the nuclear balance by insuring the survivability of the land based leg of the triad, the C3 links must be equally hardened or the advantage of survivability of the missiles will be negated.

Basing in silos with terminal defenses

Conventional, comparatively soft silos (able to withstand around 3000 psi versus on the order of 100,000 psi for superhard silos), protected by terminal defenses offer the advantage that an LCC located close to the silo could be protected by the defenses of the silo itself. In this basing mode, the vulnerability of the silo and its command and communication links with the LCC could easily be made identical. Terminal defenses would not prevent nuclear detonations in the upper atmosphere and might even use them to destroy incoming reentry vehicles. Thus the disruption of the ionosphere would be as great or greater than for surface detonation of warheads, and radio communications with the NCA might be impossible during the nuclear engagement.

The NCA could also be protected by terminal defenses and presumably radio communications could be restored after nuclear explosions stopped. The protection of ground communications between the NCA and the LCC from possible enemy attack would require defenses that attack incoming missiles in the boost or midcourse phases of their flight and would be equivalent to a complete national defense. Deep burial of an extensive ground communications link or concealment would also be expensive options. Thus it seems likely that silos protected by terminal defenses would rely on electromagnetic links to the NCA centers also protected by terminal defenses. To the extent that the NCA can be defended, this basing mode eliminates the mismatch between the survivability of the missiles and the C3 system. Targeting and damage assessment data could also be transmitted to surviving missiles by the NCA.

The weakness in this argument lies in the fact that the Soviets will probably know the location of the relatively few NCA centers and can attack them with large numbers of missiles to saturate the terminal defenses which protect them. Thus the system would still have to rely on command centers protected by mobility. Relatively ineffective defenses (say 50%) could protect a sizable fraction of the missile force from a Soviet attack but more effective defenses would be needed to assure the survival of the NCA.

A major C3 consideration for defended silos is control of the defenses themselves. Because of the short warning times available, the interceptors would have to rely on automatic warning systems. Launch of non-nuclear terminal defenses on a false alarm would not have the catastrophic consequences of launch of the missiles themselves, but such a launch might be mistaken by the enemy for an attack. At all events, waste of defensive missiles will be expensive. The vulnerability and capability of tracking and discrimination systems necessary for effective terminal defenses have been widely debated in connection with SDI technology. Large radars are expensive and will be present in limited numbers. They are soft targets and their loss can blind a defensive system. Infrared trackers are subject to "red out", or saturation by the radiation from a nuclear explosion. A major element in determining the efficiency of a defense will be the efficiency of its target acquisition, tracking and discrimination systems in a nuclear environment.

Finally the control of the defensive battle will require computer management of the defenses. The battle management software will be complex indeed and the computers which run it will have to do so in an environment of electromagnetic pulse and probably some flux of nuclear radiation. Because it takes less than a minute for a reentry vehicle to travel from the top of the atmosphere where it begins to heat up from air friction and decoys are slowed enough to be distinguishable, the tracking systems will have to operate in real time. Thus very powerful computers will be required. The defenses and the battle management software can never be tested in a real nuclear battle. Therefore bugs in the software and vulnerability of the hardware will have to be determined in a simulated environment. The C3 systems needed for controlling terminal defenses are more localized than those needed for controlling the fighting of a nuclear war but the computer systems required are of much greater complexity.

Deep underground basing

In deep underground basing, missiles are buried in tunnels in mountainsides where they are able to ride out a nuclear attack. Following the attack the missiles dig themselves out of the tunnels to the surface and launch themselves. Since deep underground basing supposes that missiles will require time following a nuclear attack to emerge from their buried silos, any retargeting instructions or final launch commands would have to reach them long after the attack. The survivability of the National Airborne Command Post (NEACP) and the other national command centers for longer than the 72 hours that the NEACP plane can stay airborne seems dubious. Thus missiles based in this mode may be given launch commands which will automatically fire the missile when it has dug itself out. Such a basing mode would be suitable for a retaliatory countervalue strike but is not viable for launching time-coordinated strikes against hardened military targets in a war-fighting scenario.

The launch commands could be transmitted to the launch control center of a missile based deep underground as for one based in a super-hardened silo and verified as it is for a regular silo-based missile. The problems of crew survival and communication with the NCA are the same as those for superhardened silo basing. Designers will have to be concerned that communications cables between the NCA and the LCC and the LCC and the launchers are as hard as the silo or the underground basing mode. Alternatively the missile could fire automatically on emergence or be fired by its crew without communicating with the NCA. In this operational mode, missiles in deep underground basing would have C3 systems like those of SLBMs.

Deep Underground Basing would make missiles less vulnerable to terrorist attack or accidents involving civilians than are silo based missiles since the missile would be difficult to remove from its tunnel in peacetime without special equipment and authorization. Accidental explosions of fuel or even detonation of the nuclear warhead would be contained by the very deep burial of the system. from its tunnel in peacetime without special equipment and authorization. Accidental explosions of fuel or even detonation of the nuclear warhead would be contained by the very deep burial of the system.

Road mobile basing

Land mobile basing modes complicate both positive and negative control of the missiles. Obviously mobile missiles cannot be connected to the NCA and launch control centers by permanent, hardened land lines. Mobile missiles could station themselves at predetermined launch points scattered across the country and connected to the NCA by hardened and buried cables but construction of such launch points would provide the Soviet targeteers with information on the probable location of the missiles and reduce the survivability that was the point of mobility in the first place. Provision of multiple launch points for each missile would cost the Soviet planners more warheads to destroy each mobile missile. Multiple launch centers would be expensive and, as the hardening of the sites increased, would approach the concept of shell game basing originally suggested by the Carter Administration for the MX.

Since communications with the NCA depend on remote transmission, it seems likely that mobile missile systems will rely on radio communications for launch instructions. The Defense Science Board task force on ICBM modernization pointed out that deployment of Midgetman in hardened mobile launchers would require development of a specialized C3 system. The task force wrote: "We believe that a direct command link to the Strategic Air Command and the National Command Authority, in addition to such intermediate launch control nodes as are necessary, would be the desired approach" (8). Presumably the miniature receivers designed for the Milstar systems, particularly for use on bombers, would function equally well for mobile missiles and the launcher could carry a deployable antenna which would allow the crew to utilize the Ground Wave Emergency Network (GWEN). Widely deployed GWEN terminals would permit the use of line-of-sight UHF communications. Airborne launch control centers could also launch the missiles using line-of-sight UHF communications.

If the Midgetman is launched remotely, there need be little concern for crew survivability during an attack except if the missile needs to make more than one dash from its position after surviving a first attack. However, if the Midgetman is to be used in war fighting after a first nuclear strike, a surviving crew will be necessary for effective retargeting and timely launch. If crews are stationed in separate launch control centers, either hardened and stationary, ground mobile or airborne, they will have to control the Midgetman by means of remote communications. The only possible exception to this would be a ground mobile LCC which would approach the Midgetman after a nuclear attack and physically attach itself to the missile. In this scenario, one LCC could launch only one missile at a given time which is certainly not efficient for war fighting.

Mobile LCCs would probably be able to control an entire squadron of missiles so that redundancy would increase the chance of survival following a nuclear attack. The survival of the crew of a LCC will pose problems for designers since men are vulnerable to fire, nuclear and thermal radiation, blast and a variety of psychological damage during the course of a nuclear attack. A barrage attack on a large area where mobile missiles are known to be based would create an environment in which crews could not function for more than a few hours because of high levels of radioactive fallout immediately following close to nuclear explosions even if the missiles themselves could escape harm, and the mobile LCCs would need protection against the effects of EMP, dust and fire.

Because mobile missiles are intrinsically more vulnerable to terrorist attacks than are silo based missiles, the missile must be launched only on receipt of a carefully guarded code. Launch will almost certainly require communications with the NCA, unlike the case of submarine launched missiles which can be launched by the crew of the boat in the event of communication loss. Thus in a crisis, commanders will be tempted to release launch codes to missile crews if the missiles are more survivable than the C3 systems linking them to the NCA. This reduces the level of command at which the decision to launch the missiles is made.

Missiles deployed in a mobile, dash-on-warning configuration will have the command and communication problems common to land mobile missiles. In addition they will depend heavily on receiving timely warning of a nuclear attack. Hobson has shown that tactical warning of a Soviet strike which would be no more than half an hour and almost certainly less, would allow the missiles too little time to dash to safety. Survival of the missiles would depend on strategic warning from the intelligence community.

In the face of a deepening crisis involving the superpowers, the President or his designate would need to decide when to instruct the missiles to dash. A sudden dash by the Midgetman force would be a clear indication to the Soviets that the US thought that a nuclear exchange was imminent. Soviet doctrine for conventional forces emphasizes the advantage that accrues to the offense which launches a preemptive and powerful first attack (9). In the face of the evidence of US preparedness presented by the dashing mobile missiles, the Soviets might decide to abandon their policy of no first use and launch a nuclear attack. Certainly elements within the military would wish to do this. Thus it is quite possible that a President might hesitate to order mobile missiles to dash until tactical warning of a Soviet strike was received. The time consumed by a Presidèntial decision following tactical warning could doom the force to destruction.

In addition to problems of positive control, land mobile missiles pose problems of negative control. Because they are mobile, these missiles are more vulnerable to both accident and terrorist activity than are silo based missiles. For this reason, Pershing mobile missiles in Europe are transported separated from their nuclear warheads that are kept in separate fixed storage bunkers. Even under these circumstances, the army expects to use 35 men to operate each Pershing II (10). These mobile missiles are deployed in battalions of four firing batteries. Each battery consists of three platoons each of which controls three transporter erector launchers (TELs). Published descriptions do not discuss provisions for crew protection during a nuclear attack.

The nuclear armed ground launched cruise missile (GLCM) is also deployed in a mobile mode in Europe. It is deployed in units of 16 missiles and four TELs controlled by two mobile LCCs which travel with 20 support vehicles and 70 personnel. Crew protection is incorporated into the design of the hardened LCC. The LCC is linked to the TEL by several communication links which include a lightweight fiber optic cable. The missiles are normally stored in protective hardened concrete shelters where the system is maintained and the warheads mated to the missile which can be stored in its cannister already fueled. The shelters are hard enough to resist physical attack and precision air strikes with conventional weapons but not to resist a nuclear strike. It is expected that under combat conditions the missiles will remove to remote locations and take advantage of local cover.(11) Both these basing modes require crisis warning to take advantage of the missile's mobility in assuring its survival. To order the GLCMs to proceed to remote deployment sites or to mate the Pershing II missiles with their nuclear warheads might have the same effect of a crisis as ordering mobile missiles to dash on strategic warning and thus suffer from the disadvantage described above.

The US has considerable experience with the security and safety of its mobile missile systems in Europe. There have been no nuclear accidents or incidents and the warheads are believed to be protected by electronic permissive actions links (PALs) which disable the warhead if the correct code is not entered in a limited number of tries. It is unlikely that the warheads would undergo nuclear detonation without the direct code to arm them. Thus in order to guarantee this negative control, the missiles must communicate with the command authorities to obtain the correct PAL codes before they can be launched. In light of the problems cited above which are known to exist with the C3 systems of the missiles in the environment of a nuclear attack, the ability of the command authorities to issue the codes once an attack is underway is by no means assured. On the other hand, a precipitous release of the PAL codes to field commanders might lead to rapid escalation of the nuclear combat in an unplanned and catastrophic manner.

In the event of a physical accident to the warhead, the conventional explosive in the warhead might detonate, strewing plutonium in the vicinity of the detonation but it is unlikely that a nuclear detonation would occur. There have been several incidents in which nuclear warheads were involved in accidents as severe as plane crashes and the resulting chemical explosions did not trigger nuclear detonations. The warheads are stored in hardened and heavily guarded bunkers which are protected from terrorist intrusions.

The missiles themselves have proved more vulnerable than their warheads. Possibly this is because less attention has been focused on protecting them. A Pershing 2 rocket motor exploded during unloading and started a fire which killed three US service men. The Army investigated the incident and discovered that the propellant used in the motor became sensitive to static electrical discharges under the right temperature and humidity. Their report added that the phenomenon was "heretofore unknown to missile scientists and the propulsion industry" (12). Such unfortunate accidents are not unknown in systems which involve highly volatile propellants but underline the need for retaining safeguards on the nuclear elements of mobile missiles in peace time even at the expense of the instant launch readiness that is obtained with silo based missiles.

In 1977, Blair and Brewer (13) pointed out that the great concern over the theft of plutonium from the civilian power industry and the subsequent manufacture of a nuclear weapon by terrorists was probably no greater than the danger that terrorists would steal a ready-made nuclear explosive. They recommended that the US take steps to secure the Minuteman force such as restricting the access of visitors to the missile silos and increasing the security systems that warn if anyone approaches the silos. The United States has taken many of these steps and is currently upgrading the radar systems that protect the perimeter of the Minuteman silos. Blair and Brewer mention the problems that are encountered with more mobile tactical nuclear weapons but note the effectiveness of the PALs which protect them.

In addition to protecting mobile missiles against accident or theft, the military must be concerned that they cannot be destroyed by commando units before they can be fired. The Army's investigation of the Pershing 2 fire turned up evidence that security at Pershing 2 sites in Germany was extremely lax. For example, the missiles were clearly visible through a wire fence and vulnerable to attack with small arms. This attracted protesters now and would be a tempting target for terrorists. The Army responded immediately by moving semi-trailer trucks to block public view of the missile and is considering more permanent shelters and other measures (14).

Recent discoveries of classified radio communication frequencies on the lists of drug smugglers raise questions of the degree to which security systems can protect codes of physical security systems from determined espionage agents. The physical security of mobile missile systems is clearly an issue which must be considered a cost in deciding whether the survivability of a mobile system justifies its economic and social costs.

Rail garrison mobile basing

In this basing mode, missiles would be mounted on railroad cars and stationed in hardened garrisons. Thus many of the negative control problems associated with mobile missiles could be prevented by tightening security around the garrison points. The garages where the missiles are garrisoned can be hardened to withstand an assault from conventional weapons or a terrorist attack relatively cheaply. Day to day operation of the missiles in garrisons would not require the enormous crews needed to service and protect missiles based in road mobile configurations and a single train could pull all the support vehicles needed to carry a crew, supplies for men and missile maintenance as well as extra fuel and communications equipment. Thus a rail mobile system requires a smaller crew than its road mobile counterpart even during the dash before a nuclear attack.

The problems of positive control of the missiles after an attack are similar to those for road mobile systems. The rail mobile missiles would still have to dash along railroad tracks on warning of an attack. Land lines could run along the railroad tracks with the tracks themselves serving as signal carriers. However the tracks can be broken by nuclear or conventional explosions so land lines would probably have to be buried and hardened against the effects of EMP. Multiple input and switching nodes would be necessary to prevent disruption of communications with the NCA by destruction of the soft input points or breaking of single rail lines.

Cross country communication lines could run to presurveyed launch points along the track but their presence would risk providing the Soviets with information on the probable location of the missile train during a nuclear barrage attack making them sitting duck targets for accurate Soviet warheads. Communications could be managed by radio as it would be easy for the train to carry an antenna which would be deployed after the missile has survived an initial Soviet attack.

Trains can carry more weight than vehicles which travel on the highways. Thus crew compartments can be hardened more easily than road mobile LCCs and the crew can carry water and other supplies to survive in a radioactive environment. Even so, protecting crews from the lethal effects of a near nuclear miss or the very high levels of radiation produced by a large number of near by surface bursts will pose a significant design challenge if it can be done at all.

Conclusions

If the major purpose of the land based leg of the triad is to provide a survivable deterrent which can respond to a Soviet first strike by attacking hardened military targets, attention must be given to improving the C3 systems to be deployed with the missiles. At the present time, there is no basing mode in which the missiles can be controlled with sufficient precision to permit retargeting and timed salvos needed to attack military targets in a war-fighting scenario. They could be launched under attack or fired by predetermined codes at preselected targets. However the C3 systems currently in place could not be used in a war-fighting mode following a nuclear attack. Basing missiles in a mobile mode increases the difficulty of maintaining communication links with the NCA in a nuclear war environment and increases problems of negative control.

If the major purpose of the land based leg of the triad is to provide a preplanned, countervalue strike against the Soviet Union following a nuclear attack, then the demands on the C3 systems of the missile are far less severe since only simple launch codes would be needed to trigger the response. At present, the C3 system might be able to provide this type of positive control of the missile.

Certainly the C3 situation for current silo based missiles could be improved, as has been acknowledged by the Reagan administration. Any situation in which the missile is substantially more survivable than the C3 systems which control it will tend to cause commanders to release launch permission codes to officers stationed in individual LCCs either during the final stages of a deepening crisis, on warning of an attack or during the first detonations of such an attack. In this case, the launch decision could be made by the relatively low ranking officers in the LCCs who would certainly be under enormous psychological stress. Such a situation is highly unsafe in a crisis and must be avoided. Attention to the survivability of positive control of a missile must match that given to survivability of the missile itself.

Mobile basing modes increase the vulnerability of the missile system to terrorist attack and to accidents. Survivability purchased by such basing modes must be weighed against both the increased vulnerability of their C3 systems and the risk they involve. Garrison rail mobile basing schemes offer better negative control over missiles than do road mobile basing schemes.

References and notes

*This work was partially supported by a Faculty Summer Research Grant from Ball State University.

1. *Report of the President's Commission on Strategic Forces*, April 1983, Reprinted by the Library of Congress,, Nov 1983.

2. Caspar W. Weinberger, *Annual Report to the Congress for Fiscal Year 1987*, Department of Defense, 5 Feb 1986, p. 214.

3. *Report of the President's Commission*, p. 10.

4. Thomas B. Cochran, William M. Arkin, and Milton M. Hoenig, *Nuclear Weapons Databook: US Nuclear Forces and Capabilities*, Ballinger Publishing Company, Cambridge, Massachusetts (1984), p. 133.

5. James M. Markham, "G.I.'s die at German base when missile catches fire," New York Times, 12 Jan 1985.

6. Art Hobson, articles in this book.

7. Office of Technology Assessment, *MX Missile Basing*, US Congress, Washington, D.C. (1981).

8. James K. Gordon, "Changes in small ICBM could delay deployment two years," Aviation Week and Space Technology, 7 Apr 1986, pp. 18-19.

9. See S. P. Ivanov, *The Initial Period of a War*, translated by the US Air Force, US Government Printing Office, Washington, D.C.(1974), particularly chapter 9, and A. A. Siderenko, *The Offensive*, translated by the US Air Force, US Government Printing Office, Washington, D.C. (1970), for a discussion of Soviet tactics.

10. Jonathan Medalia, *Small Single-Warhead Intercontinental Ballistic Missiles:Hardware, Issues, and Policy Choices*, Congressional Research Service, The Library of Congress, Report No. 83-106F.

11. *Jane's Weapon Systems 1985-86*, Jane's Publishing Company Limited, London (1985), pp. 42 and 44.

12. Richard Halloran, "Army modifying missiles to prevent mishaps," New York Times, 25 Apr 1985.

13. Bruce G. Blair and Garry D. Brewer, "The Terrorist Threat to World Nuclear Programs", Journal of Conflict Resolution, Volume 21 (1977), pp. 379-403.

14. Bill Keller, "Army terms Pershing 2 missiles vulnerable to terror in Germany," New York Times 30 Apr 1985.

Appendix A: Current C3 systems

Several recent books and articles have questioned the operation of the command, control and communications system of the US strategic forces (1). The entire C3 system of the United States is being extensively upgraded under the Reagan administration's strategic modernization program and improvements that are expected to be implemented will be described following a description of the current system.

The ultimate authority for release of nuclear weapons rests with the President of the United States and his designated successors beginning with the Vice President and continuing through the Speaker of the House of Representatives to the remainder of the cabinet in specified order. Official US policy forbids launch of nuclear weapons without clearance of the designated National Command Authority (NCA) although the President might delegate his authority in the event of a deepening international crisis.

The codes to launch nuclear weapons follow the President wherever he goes in the black briefcase known as the "football" that is carried by a military aide. These codes give the President the trigger to carry out a variety of nuclear strikes designed in advance as a part of the Single Integrated Operational Plan.

In a nuclear war-fighting scenario, the timing of weapon launches will be critical to prevent fratricide and to insure an effective strike. The type of weapon suited to the destruction of a particular type of target must be launched against that target. For example current submarine launched ballistic missiles are effective against soft targets such as industrial plants, transportation centers or airfields but they cannot reliably destroy hardened missile silos or command bunkers. Thus retargeting of weapons systems cannot be done hastily, certainly not in the 15 minutes or so that will be available following warning of a Soviet nuclear first strike. The President thus must select from one of several pre-planned nuclear attacks.

If strategic doctrine calls for a protracted nuclear combat, provision must be made for retargeting of nuclear weapons so as not to waste firepower on empty silos and in order to strike mobile Soviet nuclear weapons systems that have been held in reserve. Clearly these protracted war-fighting doctrines require much more sophisticated C3 systems than those required to launch a pre-planned retaliatory strike.

Since it is probable that Washington will be a prime target, the President and many of his designated successors may die in the early stages of the war. The NCA will be comprised of the survivors and members of the Joint Chiefs of Staff. Nuclear war will be commanded from the National Military Command Center located in the Pentagon and from the Alternate National Military Command Center located in Maryland. Neither of these command centers is hardened to resist a direct nuclear strike.

As a backup system, the United States utilizes the National Emergency Airborne Command Post (NEACP) which is an E 4B airplane instrumented for communication with field commanders and stationed on alert at Grissom Air Force Base in Indiana. In addition, an aircraft known as Looking Glass is kept permanently in the air flying patterns over the midwest carrying a SAC general on board. NEACP presents a difficult target for Soviet strategic systems as long as it has sufficient warning to get airborne before Grissom is hit. Unfortunately while NEACP is a highly mobile system, it can communicate with the rest of the US strategic command system only through 14 fixed ground centers which are relatively soft targets for nuclear weapons. NEACP cannot directly launch any nuclear weapon. Looking Glass is capable of communicating with the launch control centers which actually fire silo-based missiles but NEACP does not have this capability. The aircraft are vulnerable to nuclear blast and dust as was convincingly demonstrated when commercial aircraft of similar designs experienced engine trouble in the dust cloud generated by the eruption of Mt. St. Helen. Both Looking Glass and NEACP have limited flight duration (72 hours for the E 4B planes with aerial refuelling). Thus, as eloquently pointed out by Steinbruner in 1981 (2), the survival of the NCA during a protracted nuclear engagement seems dubious at best.

In the event that the NCA survives, it must still communicate launch orders, euphemistically known as an Emergency Action Message (EAM), to silo-based missiles. According to Arkin and Fieldhouse (1), the EAM is passed to one of the military control centers where it is verified and recoded as a Nuclear Control Order which is passed to the military commands such as SAC, the strategic air command, or EUCOM, the unified command in Europe. The launch order is passed down the military chain of command to the actual launch control of the weapon. At each stage of the chain of command, there are elaborate procedures to verify that a launch order is real. The "two man rule" implies that each stage requires the independent consent of at least two men acting in consort to launch the missile.

In the case of submarine launched missiles, the co-operation of more than two members of the crew is required since submarine launched missiles must be capable of launch without communication with the mainland if they are to provide a retaliatory strike against the Soviet Union in the event of nuclear attack on the continental United States. Careful design precludes the possibility that a single crazed crew member could force a launch at gunpoint. Tactical nuclear weapons based in Europe are protected by Permissive Action Links (PALs) which are electronic codes needed to arm the warheads and which are possessed only by the highest command authorities. If the wrong code is entered into the weapon for some limited number of times, the warhead will destroy itself.

At the present time, US intercontinental ballistic missiles, the Minuteman and the MX, are based in hardened silos connected to launch control centers (LCCs) by buried cables. The Minuteman is based in flights of 10 missiles with an associated buried LCC which is separated from the missiles it controls by distances on the order of miles. Missile silos are separated from each other by distances of 6 kilometers or so in order to increase survivability in case of attack.

Each squadron of 50 missiles thus has 5 LCCs associated with it and connected to it by hardened underground cables. Each LCC is manned by two carefully selected Air Force officers. If a NCO is received by the LCC, the officers separately decode it and validate it. According to Ford, the codes and validation procedures are kept in a safe which can be opened only by two keys, one of which is carried by each officer. If the message is in any way defective, the officers will not launch until they are assured by command authorities that it is valid. Otherwise they will procede to launch their missiles by typing a six digit code contained in the NCO into the positive enabling switch in their LCC.

If one LCC tries to launch without a valid order, it can be overridden by the other four centers attached to the squadron. In each LCC, the crew must turn two keys simultaneously in locks which are far enough apart so that one man cannot turn both keys. Two LCCs must do this simultaneously. Thus it is hopefully impossible to launch a missile without authorization in peacetime conditions. In the event of a nuclear attack, one surviving LCC can launch all 50 missiles in the squadron although there will be a delay of half an hour or so in the launch according to Blair. The LCCs associated with one squadron are not connected to the missiles of any other squadron.

In the event that all LCCs of a squadron are destroyed in a nuclear attack, Looking Glass and other airborne SAC command posts can launch the missile force by transmitting launch codes at ultrahigh frequency to an antenna located next to each silo. Alternatively a set of Minuteman missiles stationed at Whiteman Air Force Base carry transmitters instead of warheads. On launch, this Emergency Rocket Communication System will transmit launch orders at two UHF frequencies for about half an hour. The orders can fire the Minuteman force and can be changed before launch of the transmitting missiles by commanders in Looking Glass. The NEACP plane cannot launch the missile force except by communicating with one of the national military command centers or Looking Glass. If the LCCs are not destroyed, Looking Glass must communicate with the surviving center to launch the missiles in that squadron.

The missiles cannot be launched without the correct six digit code which is not known to the officers in the LCCs. Thus loss of the command and control network can neutralize the silo-based missile force. Retargeting to strike time urgent targets as is necessary in nuclear war fighting scenarios clearly requires even better communications between the LCCs and the military commanders for transmission of new targeting data to the missiles.

The environment following a nuclear attack is not likely to be conducive to the effective operations of communication systems (3). In the first place, blast and thermal radiation would destroy targets located on the ground. Unhardened communications centers are easily destroyed by airbursts some distance away and antennas are notoriously soft targets since they are damaged by high winds and flying debris. Buried cables are generally destroyed only when they are cut by a nuclear blast. Aircraft caught on the ground during nuclear strikes are extremely soft targets. In the air, they can be destroyed by close nuclear detonations although this would stress Soviet trackers since the flight times of even submarine-launched ballistic missiles launched just off the US coast to the center of the US are on the order of 10 minutes and the aircraft travel in random patterns at speeds on the order of 500 mph. Looking Glass might be vulnerable to a first strike attack if it flew patterns that were sufficiently reproducible so that Soviet trackers could accurately predict its location. Thus buried silos are much more difficult to destroy than the NMCC or the land links for communication with NEACP, which are nothing more than AT&T switching centers.

Warning times on the order of 15 minutes are required to launch airborne command centers. It is for this reason that NEACP is now based in Indiana rather than in Washington since Soviet submarine-launched missiles could reach Washington with far less warning time than 15 minutes. If the civilian chain of command were to be maintained, national leaders would have to leave Washington for Indiana as a crisis developed to the flash point in order to reach NEACP before an attack on Washington. If the President or his representative does not reach NEACP or another surviving command post, the military chain of command would take over conduct of the war.

Once in the air or surviving buried on the ground, the command centers must communicate with the weapons systems by radio or buried cables. The communications network which will control a nuclear war is part of the World Wide Military Command and Control System which links all components of the US military in peace and war. Trailing antennas and complex digital circuits are vulnerable to the effects of Electromagnetic Pulse (EMP), a high voltage pulse produced over large areas by the interaction of nuclear warheads detonated at high altitude with the atmosphere. Power lines and long distance cables are particularly vulnerable to being burned out by EMP. Solid state electronic components are damaged by the direct interaction of nuclear radiation with the materials which comprise them. These effects are called Transient Radiation Effects on Electronics (TREE).

The susceptibility of an electronic device to TREE depends on the detailed design of the device although in general, devices which are more densely integrated are more vulnerable. Gallium Arsenide chips are generally more resistant to radiation damage than are conventional doped silicon devices. TREE can also damage wiring in a circuit whose major components are left intact. Thus even a near miss which leaves a command center physically intact can render it inoperable by destroying the circuitry, antennas or cables by which it can communicate with the outside world. Crews manning command centers may be killed by heat. blast or radiation which would not damage their equipment. The psychological state of command center crews during a nuclear exchange is difficult to predict and may severely limit their efficiency during a crisis.

If land lines cannot be used for communication, messages must be sent using some portion of the electromagnetic spectrum. If the sender and receiver are not within line of sight of one another, messages must be relayed by bouncing off the ionosphere or off a satellite. Nuclear explosions disrupt the atmosphere and the effects on communications depend on the region of the electromagnetic spectrum being used. In particular the effects of large yield, high altitude nuclear bursts can spread over many miles. Low frequency, very low frequency or extremely low frequency radio communications (3-300 KHz) are the least affected of the radio frequencies used for strategic communications. For periods on the order of minutes and up to hours, signals at these frequencies may be distorted both through phase shifts and fluctuations in strength. The signals are long range and the disruptions do not necessarily interrupt all transmission of data for extended periods although they may seriously distort messages. Unfortunately these radio frequencies also require the use of very large antennas and receivers which are vulnerable to both physical destruction and EMP. Data rates at these frequencies are very low and thus they serve mainly to transmit very short coded orders triggering prearranged attack plans and are not very useful for sending detailed launch time and target information.

The high frequency band (3-30 MHz) requires smaller antennas and lighter equipment to support higher data rates transmitted with much less power than messages at lower frequencies. For long range communications, HF signals bounce off the ionosphere. Nuclear detonations disrupt the ionosphere so that it absorbs rather than reflects HF signals. Large areas may be blacked out for times on the order of tens of hours. Very high frequency signals (30-300 MHz) are short range but offer high data rates with small equipment. Even without the effects of a nuclear war, the range limitation is severe, as is familiar to listeners to FM radio whose signals are transmitted in this frequency range.

Long range transmission of VHF signals uses reflection off the ionosphere which introduces the vulnerability of HF systems although the effects at VHF frequencies are generally shorter lived. Ultra high frequencies (300-3000 MHz) are not disrupted by nuclear effects except when a fireball is present between the sender and receiver. Without satellite relay, UHF signals travel only over line-of-sight paths which limit communications to 200 miles range between aircraft and ground terminals and 300-450 miles between aircraft. SAC bombers will be controlled by a link of command airplanes stationed in flight paths that keep them linked by UHF communications. UHF equipment is compact, low-power and operates at high data rates.

For long range communications, UHF signals can be retransmitted from satellites as can signals at higher frequencies such as laser wavelengths, super (3-30 GHz) or extremely (30-300 GHz) high frequencies. Such communications systems offer high data rates and small antennas. They may be blacked out by nuclear effects on the atmosphere for times up to hours. UHF antennas are not very directional and are therefore especially vulnerable to jamming.

More significantly, these communication links depend on satellites for success and are therefore vulnerable to the anti-satellite (ASAT) weapons now being developed (4). At the present time, tested ASAT systems are effective against satellites in low orbit only but there is reason to assume that the technology will mature to make them effective against satellites in higher orbits. The weapons designed by SDI for attacking incoming intercontinental ballistic missiles will be effective ASATs. Not only communication satellites will be vulnerable but also the vital early warning and reconnaissance satellites, satellite systems designed for damage assessment during a nuclear exchange and navigational satellites used in guiding some missiles.

In addition to effects on equipment, nuclear detonations may kill men or render them psychologically incapable of performing their jobs. An enemy may also jam communications systems or attack them with antiradiation conventional weapons such as the US HARM (AGM-88) missile. Digital communications are particularly vulnerable to spoofing or transmission of false messages by an enemy since digital data look the same in any language.

Appendix B: C3 modernization program

The Reagan administration has recognized the vulnerability of US C3 systems. The publication United States Military Posture FY 1987 produced by the Joint Chiefs of Staff states: "Command and control systems must be as survivable and enduring as the forces they support..." (5). Secretary of Defense Weinberger writes: "Deterrence requires that strategic C3 systems be sufficiently effective, reliable, survivable, and redundant to ensure that the National Command Authorities (NCA) receive timely warning of an attack and have an assured means of passing retaliatory orders to our strategic forces" (6). The administration has undertaken a large scale program to upgrade our national C3 systems in three major areas: surveillance, warning and damage assessment capability; command of nuclear weapons; and survivable strategic communications.

In the area of surveillance, warning and damage assessment, the strategic modernization program has stressed providing detailed reliable early warning of a Soviet strike and a survivable capability for assessing damage during a protracted nuclear exchange. Early warning efforts have concentrated on detection of attacks from low flying cruise missiles and early warning of submarine launched ballistic missiles. Program elements include upgrades of the Pave Paws phased array radar warning system for submarine launched ballistic missiles (6,7) and construction of two new sites in Texas and Georgia to scan the southern borders of the United States.

The Ballistic Missile Early Warning System (BMEWS) network of radars in Thule,Greenland, Clear,Alaska and Fylingdale, England are being replaced with phased array radars to increase their tracking capability, range and angle of coverage while decreasing their power consumption (8). The BMEWS upgrade will be supplemented by a phased-array warning radar sited in North Dakota called PARCS, the Perimeter Acquisition Radar Attack Characterization System (6), and by Seek Igloo, a network of 13 long range radars in Alaska (8).

The 31 aging radars of the Distant Early Warning System (DEW Line) are being replaced by 52 more modern radars to form the NWS (North Warning System) Network (8). To warn of the arrival of low-flying cruise missiles and aircraft and to supplement Pave Paws, the Air Force has begun to install over-the-horizon backscatter (OTH-B) radars which utilize a high frequency beam that travels beyond line-of-sight range by scattering off the ionosphere. The frequent disturbances of the ionosphere in polar regions render OTH-B radars unreliable in these regions but the system is expected to be effective at lower latitudes (8).

Finally the GEODSS (Ground-based Electro-Optical Deep Space Surveillance) system uses five equatorial sites spaced approximately evenly around the world to detect space objects out to geosynchronous ranges (9).

Satellite upgrades include new microwave sensors for the Defense Meteorological Satellite Program and hardening of the satellites to protect against the effects of radiation. New EHF communications links with the satellites will make them more useful following a nuclear strike (10). Defense Support Program (DSP) warning satellites will be linked by laser communications so that the satellites will not be totally dependent on the ground station in Australia (8). Old DSP satellites will be kept in orbit as spares in the event of nuclear attack and new models are equipped with "upset recovery" to restart electronic circuits disrupted by the effects of a nuclear blast (10). Damage assessment capabilities will be improved by mounting sensors of the Nuclear Detonation Detection System on the satellites of the Global Positioning System (6). Such sensors would locate the sites of nuclear detonations but satellite imagery or inspection from aircraft would be required to assess the extent of the damage produced by any detonation.

To ensure command of the nuclear forces and the ability to use the data collected by the new sensors in fighting a nuclear war, the hardened headquarters of the North American Aerospace Defense Command in Cheyenne Mountain is being further hardened against the effects of nuclear detonations and equipped with modern data handling facilities. A new facility dealing with satellite control and data from space, the Space Defense Operations Center, is being installed in the mountain (9). The newly formed Space Command is requesting back up facilities to support the vulnerable satellite control center that is currently in California but which will soon be moved to the Space Defense Operations Center (11) and attention is being given to the vulnerable ground link stations that are essential to satellite communications. No final solution has been found to the problem of protecting the ground stations but suggestions include placing them in hardened airborne command posts, mobile land posts or simply proliferating them as is being done with the Global Positioning System (Navstar GPS) (10). The fleet of four E-4B aircraft that constitute NEACP are being hardened against EMP and their communication systems are being upgraded. Suggestions for the development of Ground Mobile Command Posts (or even Sea Mobile Command Posts (12)) appear frequently in the literature (12,13) but at this writing appear to be speculative solutions to the problem of survivable command centers.

Communication systems are being modified to survive a nuclear strike in several ways. The Ground Wave Emergency Network (GWEN) is a land-based communications network hardened against EMP effects and providing many switching nodes so that destruction of a few switching centers cannot disable the entire network. GWEN uses line-of-sight UHF radio communications and has a very low data rate (75-100 bits/s) and thus is suitable for short emergency messages and not for complex retargeting information. New nodes are gradually being added to the system which will be completed by the end of the decade (6,9). GWEN will be connected to operating SAC bases by hardened cables and the Aircraft Alerting Communications EMP protection system will field EMP detectors to let base commanders know that emergency communications should be used (9). Nodes connecting GWEN with satellite communications systems and NEACP and Looking Glass will be limited in numbers and may form tempting targets for enemy attack.

The Milstar (military strategic, tactical and relay) satellite communication network is certainly the most expensive component of the C3 modernization program (13). Milstar will consist of a constellation of nine satellite communication relays, six in geosynchronous orbit and three in elliptical polar orbits. Because they will receive at 44 GHz and transmit at 20 GHz, ground stations can use very small antennas and thus be made portable and mobile. Four thousand ground terminals will eventually be deployed with all branches of the military. These EHF band communications are resistant to the effects of nuclear explosions and difficult to jam (14). Milstar will eventually replace the current Air Force Satellite Communications System and the Defense Satellite Communications System although these satellites are still being replaced by newer models that are more resistant to nuclear effects (6). Milstar communications links carry peacetime operating orders and in war will carry tactical battle commands for nuclear and conventional weapons as well as reports of damage to weapons systems, targeting data and launch commands for strategic nuclear weapons.

The Command and Communications systems of silo based Minuteman missiles have also been updated, partially in preparation for the deployment of the MX Peacekeeper missile in these silos. Specifically new hardened cables have been strung between silos and the LCCs. LCC facilities will be modified to support a new integrated C3 system which will eventually connect 10 rather than the current 5 LCCs. Displays and controls are being improved and the data handling systems are being upgraded to allow launch more quickly following receipt of a launch command. A new radar system is being installed to protect the silos from physical intrusion during peacetime and is expected to reduce the numbers of false alarms generated by the system. Finally the power system of the missiles is being reworked to provide longer and more assured survival following a nuclear attack. Batteries are being replaced by high energy lithium batteries which last well and are resistant to nuclear effects (15).

Clearly these modernization programs represent a major effort to design a sophisticated C3 system that will survive the effects of a nuclear exchange. Whether the proposed systems provide such a capability is the subject of many lengthy debates.

References and notes to the appendix

1. Paul Bracken, *The Command and Control of Nuclear Forces*, Yale University Press, New Haven (1983); Bruce G. Blair, *Strategic Command and Control: Redefining the Nuclear Threat*; The Brookings Institution, Washington, D.C. (1985); Ashton B. Carter, John D. Steinbrunner, and Charles A. Zraket, editors, *Managing Nuclear Operations*, The Brookings Institution, Washington, D.C. (1987); Daniel Ford, *The Button: The Pentagon's Strategic Command and Control System*, Simon and Schuster, New York (1985); Desmond Ball, "Can nuclear war be controlled?," Adelphi Papers, No. 169, The International Institute for Strategic Studies, London (1981); William M. Arkin and Richard Fieldhouse, "Nuclear weapon command, control and communications," *SIPRI Yearbook 1984*, Stockholm International Peace Research Institute (1984).

2. John D. Steinbruner, "Nuclear decapitation", Foreign Policy, Winter 1981-82, pp. 16-28.

3. For discussions of the effects of nuclear detonations on communications systems see the following: Desmond Ball, op. cit.; Michael King and Paul Fleming, "An overview of the effects of nuclear weapons on communications capabilities", Signal, Volume 34, 1980, pp. 59-66; Office of Technology Assessment, *MX Missile Basing*, US Congress, Washington, D.C. (1981), Chapter 10; Samuel Glasstone and Philip J. Dolan, *The Effects of Nuclear Weapons*, Third Edition, US Department of Defense and the Energy Research and Development Administration, Washington, D.C. (1977).

4. See, for example, Bruno Pattan, "Hunter-killer space vehicles," Journal of Electronic Defense, Volume 8, 1985, pp. 49-54.

5. *United States Military Posture FY 1987*, Organization of the Joint Chiefs of Staff, Washington, D.C. (1986), p 9.

6. Caspar W. Weinberger, *Annual Report to the Congress for Fiscal Year 1987*, Department of Defense, 5 Feb 1986, p. 223-224.

7. David J. Lynch, "Pave Paws to get blue suits," Defense Week, 17 Mar 1986, p. 10.

8. "Mending America's electronic fences" and "Military space and SDI spending climbs," Critical Program Reviews, Defense Electronics, Volume 17, Dec 1985, pp. 49 ff.

9. John Haystead, "AF C3I procurement centers," Journal of Electronic Defense, Apr 1986, pp. 28 ff.

10. Jack Cushman, "AF starts satellite improvement," Defense Week 17 Mar 1986, p. 5.

11. "Space command wants backup command centers," Defense Electronics, Volume 18, Aug 1986, p. 28.

12. Alan J.Vick, "Post-attack command and control survival: options for the future", Orbis, 1985, pp. 95-117.

13. *Strategic Command, Control and Communications: Alternative Approaches for Modernization*, Congressional Budget Office, Washington, D.C. (1981).

14. "Critical C3 programs emphasize survivability, jam resistance," Critical Program Reviews, Defense Electronics, Volume 18, Jul 1986, pp. 91 ff.

15. *Jane's Weapon Systems 1985-86*, Jane's Publishing Company Limited, London (1985), p. 19.

Part V Chapter 7

Survivability of mobile Midgetman

Art Hobson

Introduction

Three Midgetman basing modes have gotten serious DOD consideration (1-4): hardened mobile launchers (HMLs) moving randomly over military land (*random mobile*); HMLs parked at Minuteman silos ready to dash on warning onto surrounding roads and farms (*dash mobile*); and superhard silos. I study the two mobile modes in this chapter, and superhard silos in Chapter 11.

The missile, its mobile launcher, and the two mobile basing schemes are described in Part IV Chapter 5 of this study, and also in (4-11); see Figures 1 and 2.

Barrage attack against random mobile

Random mobile deployment restricts the HMLs to a maximum of about 30,000 km^2 of useable DOD land. However, current plans specify peacetime dispersal over only 10,000 km^2 situated on the *perimeters* of perhaps five tracts of this land (7,12). "During periods of increased tension," the HMLs would expand their operating area to 20,000 km^2 by filling the *interiors* of these tracts (12). "In the event of imminent attack" (5) the HMLs would dash outward off of DOD land onto adjacent roads and public land, expanding the operating area still further. We will call these three conditions *peacetime*, *alert*, and *attack*.

Figure 1. The hardened mobile launcher (HML).(Photo courtesy of US Department of Defense.)

The HMLs would reside nearly permanently in the field, remaining stationary most of the time but moving often enough to defeat any Soviet attempt to target their individual locations.

One can think of several ways to attack mobile missiles: sabotage, short-warning targeted attack against observed HML locations, electronically directed maneuvering warheads that can chase down individual HMLs (launched either on missiles or by airplanes), attacks on the missile's command and control system, and a general barrage of the entire operating area. Sabotage can presumably be defeated, especially on the DOD deployment areas, by sufficiently thorough security precautions. The targeted attack can be defeated by moving the HMLs randomly and often: this is precisely the point of mobility. Directed ICBM warheads will probably not be technically feasible for the Soviets by the mid-1990s (13-15); we discuss this question further below. Command and control vulnerability is discussed by Ruth Howes in Chapter 6. The primary threat against land-mobile missiles, and the threat against which the system is primarily designed, is the area barrage. I now turn to the analysis of this threat.

Figure 2. Hardened mobile launcher, top view.(Photo courtesy of the US Department of Defense.)

A groundburst of yield y will destroy any HML of strength S within a "circle of destruction" of radius RD found from Chapter 3 Eq. (5) (unapproximated form). But for lower-strength targets such as the 2 atm HMLs, an optimum-height airburst is significantly more destructive than a groundburst (16,17). Against an HML, the airburst RD is 10% larger than the groundburst value, so that the area of the circle of destruction is 21% larger.

The barrage pattern that destroys the largest fraction of the HMLs, while at the same time not "wasting" any attacking warheads by allowing the circles of destruction to overlap, is the hexagonal close-packed pattern shown in Figure 3. Straightforward geometrical calculation shows that, for this pattern, the fractional area covered by circles of destruction (the "efficiency") is 91%, and the number of attacking warheads needed per unit area is

$$N/A = 0.289 / RD^2 \qquad \text{(efficiency = 91\%)} \qquad (1)$$

Other non-overlapping patterns are less efficient. More efficient patterns would require more warheads because their circles of destruction overlap. For example, an overlapping hexagonal pattern having 100% efficiency (no "survival gaps" between circles of destruction) would require 33% more wareads to barrage a given area.

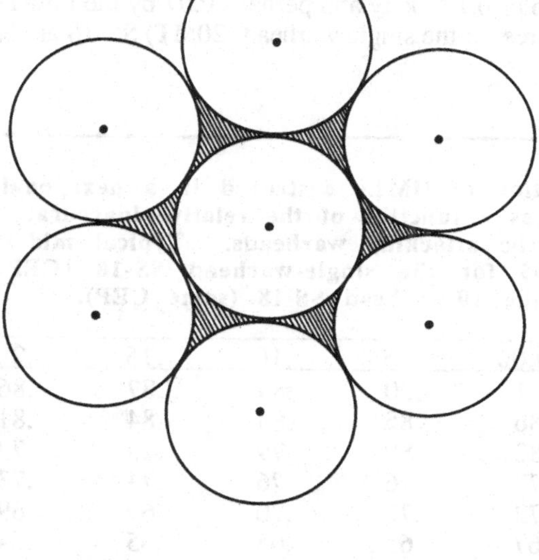

Figure 3. The hexagonal close-packed barrage pattern. Theoretically, launchers located in the shaded regions, between the circles of destruction, would survive.

In this paper, I will assume that the barrage pattern of Figure 3 would be used. However, the actual destruction fraction will deviate from the 91% geometrical coverage of Figure 3, for various reasons. As in silo attacks, the main reasons are inaccuracy and unreliability. Inaccuracy will randomly displace the centers of the circles of destruction by a distance on the order of the CEP (the CEP is a measure of the attacking warhead's inaccuracy--see Chapter 3), while unreliability ("duds") will leave some circles uncovered. Inaccuracy and unreliability are both favorable to the US, since they reduce the efficiency. On the unfavorable side, some of those HMLs that happen to reside in the "gaps" between the circles of destruction (the shaded areas in Figure 3) would surely be destroyed, because these would feel the cumulative effect of three near-threshold blasts.

Since CEP< < RD for relatively soft targets such as HMLs, we expect the inaccuracy effect to be small. Furthermore, this effect can be made smaller by using larger attacking warheads (e.g. single-warhead SS-18s instead of 10-warhead SS-18s) having the same CEP. As we will show later, this de-MIRVing option does not affect the number of *missiles* needed to barrage a given area.

Computer calculations (18) show that , for realistic values of CEP and RD, inaccuracy has indeed only a small effect, reducing the destruction fraction to 87-90%, only slightly below the ideal (zero-CEP) value of 91%. Unreliability has a more important effect. Calculated efficiencies, as a function of the ratio CEP/RD and of the reliability R, are shown in Table 1. The 10-warhead SS-18 has a CEP/RD ratio of 0.17 today and perhaps 0.07 by the mid-1990s, while the corresponding Figures for the single warhead (20MT) SS-18 are 0.05 and 0.02.

Table 1.* Fraction of HMLs destroyed in a hexagonal close-packed barrage pattern, as a function of the relative inaccuracy CEP/RD and reliability R of the attacking warheads. Typical mid-1990s CEP/RD ratios are .02-.05 for the single-warhead SS-18 (CEP=.10-.25 km), and .07-.17 for the 10-warhead SS-18 (same CEP).

CEP/RD =	0.00	.05	.10	.15	.20	.25
R = 100%	.91	.90	.89	.87	.86	.85
95%	.86	.85	.84	.84	.81	.80
90%	.82	.80	.79	.79	.77	.76
85%	.77	.76	.76	.74	.73	.72
80%	.72	.72	.70	.69	.69	.68
75%	.67	.65	.65	.65	.64	.63
70%	.65	.62	.61	.61	.61	.60
65%	.62	.59	.58	.56	.56	.55

* I thank Herbert Nelson of the US Naval Research Lab for these calculations.

At the pessimistic end of the "vulnerability error bars" (the range of plausible vulnerabilities--see Chapter 3), we assume 90% reliability and low CEPs. Table 1 then shows an efficiency of about 80%. However, half of the surviving 20% will be in the "survival gaps" shown in Figure 3, and will feel two or three near-threshold blasts; at the centers of the gap (Figure 3), the overpressure from each of the three blasts is calculated to be 1.3 atm (20 psi). A pessimistic assessment must conclude that 2 atm HMLs might be destroyed by this rapid-fire pounding, so that the barrage efficiency becomes 90%.

At the optimistic end, we assume CEP/RD = 0.17 and reliability = 70%. Table 1 then shows an efficiency of 61%.

Thus our vulnerability error bars run from 61% to 90%. Plausible intermediate assumptions are: CEP/RD = .10, R = 80%, 50% of the HMLs residing in the "survival gaps" are actually destroyed by the two or three near-threshold blasts. With these intermediate assumptions, the survival rate (efficiency) is 75%.

MIRVing the attack makes no difference

What effect does increasing the number of warheads on the attacking missiles have on a barrage using a fixed number of missiles? We will find, perhaps surprisingly, that MIRVing makes no difference.

We begin by asking how many smaller MIRVed warheads of yield y could be carried by a single-warhead missile of yield Y. This number is called the "fractionation number." One might guess that the fractionation number is $f = Y/y$, because the MIRVed missile might be expected to carry the same total yield Y. However, the actual number is less than this, because the MIRVed warheads don't fill the entire volume that is available to the un-MIRVed warhead (Figure 4). Empirically, the formula

$$f = (Y/y)^{2/3} \qquad (2)$$

works fairly well (19). For example, the single-warhead SS-18 Mod 3 has Y= 20 MT, while the SS-18 Mod 4 (the same missile, MIRVed) has y = 0.5 MT. Thus Eq. (2) predicts f = 11.8, close to the actual number f = 10.

Figure 4 shows the probable reason for the fractionation rule of Eq.(2): In an f-MIRVed missile, f reentry vehicles (RVs) of yield y are crowded into a missile which, in single-RV form, would carry yield Y. Let the missile's radius be R. The f conical RVs (i.e. the heat shields surrounding the explosive material) must be crowded together onto the base area πR^2 that fits the single warhead of yield Y. Figure 4 confirms this crucial point. Thus, letting r be the radius of the small RVs, $f\pi r^2 \approx \pi R^2$, so $f \approx R^2/r^2$. But, assuming that the large and small RVs (yields Y and y) have the same conical shape (same cone angle), these RVs have volumes whose ratio is $V/v = R^3/r^3$, thus $f \approx (V/v)^{2/3}$. Making the plausible assumption that yield is proportional to warhead volume, we get Eq. (2).

Note added in proof: The Department of Defense has recently restricted the publication of this photograph, so we are unable to reproduce it here. However, photos of MIRV payloads, confirming the point made on the preceding page, are widely available in the literature. See, for example, p. 75 of ref. 15, or Dietrich Schroeer, *Science, Technology, and the Nuclear Arms Race*, John Wiley & Sons, New York (1984), p. 150, or John Turner and SIPRI, *Arms in the '80s*, Taylor & Francis, London (1985), p.32.

Figure 4. A mockup of the MIRV payload for the MX missile. Note that the base of each warhead must fit onto the same circular area.

It is worth noting that, if the missile's nose cone were long enough, or the individual warheads short enough, that two tiers could be stacked inside the nose cone instead of the single tier shown in Figure 4, the total "floor" area available for stacking would be increased by some fraction α (between 0 and 1), and Eq. (2) would be replaced by $f=(1+\alpha)(Y/y)^{2/3}$.

Using Eq. (1), the number of *missiles* of fractionation number f needed to barrage area A is

$$M = N/f = A(1/f)0.289/RD^2, \tag{3}$$

where N is the number of *warheads*. But Eq. (2) implies $f \propto y^{-2/3}$, while Chapter 3 Eq. (5) implies $RD^2 \propto y^{2/3}$. Thus y cancels in Eq. (3), i.e. MIRVing makes no difference.

One caveat: MIRVing makes no difference so long as blast overpressure is the dominant destruction mechanism. But at very large fractionations, i.e. at smaller warhead yields, radiation effects become the dominant mechanism against HMLs strengthened to 2 atm. Depending on the amount of radiation shielding present in deployed HMLs (unclassified details are not available), radiation effects might begin to dominate at yields of 200 kilotons (for high shielding) to 500 kilotons (for low shielding). To take advantage of this radiation-dominance, the Soviets would need to have sufficient accuracy to lay down a decent barrage pattern using smaller yields around 200 kilotons (20). For further details, see Chapter 10.

How many missiles are needed for barrage?

As an important example, we calculate the number of SS-18 *missiles* needed per km^2 of operating area to barrage HMLs hardened to 2 atm. Since MIRVing makes no difference, we can represent SS-18s in Eq. (3) with f = 1, Y = 20 MT. For such a missile, attacking a 2 atm HML, Chapter 3 Eq. (5) predicts a groundburst RD of 5.00 km. The optimum-height airburst RD is 10% larger (16,17), or 5.50 km. Thus Eq. (3) implies that the SS-18's barrage effectiveness is $M/A = 0.289/RD^2 = .0095\ km^{-2}$, or 9.5 SS-18s per 1000 km^2. As shown previously, this barrage will destroy 61-90% (75% is a plausible estimate) of the deployed HMLs within the barraged area.

The attack would come from SLBMs, not ICBMs

Although the above calculation assumed an attack by SS-18 ICBMs, this missile would probably not be used for a barrage attack on HMLs. Indeed, ICBMs might not be used at all against HMLs. For one thing, the SS-18 is more accurate than needed to lay down an effective barrage pattern. The Soviets would probably save their SS-18s for tasks requiring high accuracy, such as the destruction of hardened silos.

More importantly, the SS-18 and all other ICBMs take about 30 minutes to reach their targets, giving plenty of time for the HMLs to spread out over a very large deployment area. *Attacks on mobile missiles would be more effective if they came from off-shore, short-warning SLBMs*, not ICBMs. Attacks on mobile missiles are similar to attacks on bombers: time, not accuracy, is of the essence. As with bombers, the main threat against mobile missiles is off-shore submarines, not ICBMs.

Thus we need to calculate the barrage effectiveness of other attacking missiles, especially SLBMs. Rather than calculating the effectiveness of other missiles directly, it is more illuminating to calculate their "SS-18-equivalence." For example, the land-based SS-24 is reported to have f = 8, y = 0.5 MT (21). Thus Eq. (2) predicts that the SS-24's barrage effectiveness is equivalent to that of a single warhead of yield $Y = yf^{3/2} = 11.3$ MT. According to Eq. (3) and Chapter 3 Eq. (5), the barrage effectiveness of two missiles can be compared by comparing their $Y^{2/3}$ values. Thus one SS-24 has the barrage effectiveness of $(11.3/20)^{2/3} = 0.68$ SS-18s. In this manner, we calculate the barrage effectiveness of various Soviet missiles in "SS-18 units" (Table 2). Table 2 also shows the single-warhead yield Y that the missile could carry if it were un-MIRVed, the throwweight (the weight launched into space after the booster rockets have fallen off), and the Y/throwweight ratio. Although throwweight is often used to measure barrage effectiveness, Table 2 shows that it is not the best measure, because some missiles are more efficient carriers of megatonnage than others, per tonne of throwweight. For example, the S-24 carries only 35% of the SS-18's throwweight, but has 55% of its yield and $0.55^{2/3} = 67\%$ of its barrage effectiveness.

The barrage-equivalence of a modern Soviet SLBM, either the SS-N-20 or the SS-N-23, is calculated to be 0.33 SS-18s. Thus about 29 SLBMs (3 x 9.6) are needed per 1000 km^2 to barrage HMLs. Furthermore, the efficiency of the SLBM attack would be less than that of an SS-18 attack, due to the reduced CEP/RD ratio (Table 1).

Despite the fact that the barrage might come from SLBMs, we will continue our calculations in terms of SS-18-equivalents, because SS-18 data is fairly well established while SLBM data is tenuous.

Table 2. Barrage effectiveness of several Soviet missiles, in SS-18-equivalents, and other Soviet missile data.

	warhead (f x y) (MT)	equiv yield Y (MT)	barrage effect (SS-18s)	throw-weight (tonnes)	Y/throw-weight (MT/tonne)
SS-17	4 x .75	6	.45	3	2.0
SS-18 mod 3	1 x 20.0	20	1.00	8	2.5
SS-18 mod 4	10 x .5	20	1.00	8	2.5
SS-19 mod 2	1 x 8.0	8	.55	4	2.0
SS-19 mod 3	6 x .55	8	.55	4	2.0
SS-24	8 x .5	11	.67	2.75	4.0
SS-25	1 x .5	.5	.085	1.5	.33
SS-N-20	7 x .25	3.7	.33	3	1.2
SS-N-23	10 x .25	3.7	.33	NA	NA

Random-mobile vulnerability

It is now straightforward to calculate the Soviet attack required to barrage various operating areas. It is most instructive to study the operating area as a function of dash time. We look at both the case of a dash beginning from the 10,000 km^2 peacetime operating area, and from the 20,000 km^2 alert operating area. The alert area is reported (7,12) to consist of five separate regions. Not knowing the precise shapes of these regions, we will assume that they are simply circles (Figure 5). The alert area is then five circles, each R_o = 36 km in radius, and the peacetime area is five rings extending from r_o = 25 km to R_o = 36 km along the perimeter of these circles.

Although the top design speed of the HML is 80 km/hr (50 mph) on roads, the average speed obtained from actual tests over 110 km of plausible deployment terrain is only 47 km/hr (6). Thus we assume a 47 km/hr dash speed.

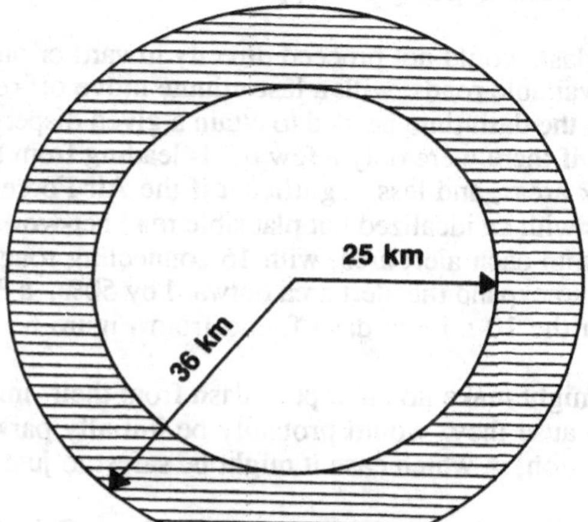

Figure 5. The peacetime and alert random-mobile operating areas assumed in this chapter. The Figure shows only one of the five operating areas. The peacetime area, on the perimeter, is shaded. The alert area includes the peacetime area and the inner circle. Upon alert, HMLs would would move inward to cover the entire alert area. "In the event of imminent attack" HMLs would dash outward from the alert area.

With these assumptions, we calculate the two random-mobile graphs of Figure 6. In addition to giving the deployment area as a function of dash time, Figure 6 gives the number of SS-18s needed to barrage this area, using 9.5 SS-18s per 1000 km^2. As discussed above, this barrage would destroy 61-90% (a plausible value is 75%) of the HMLs in the barraged region.

Figure 6 is based on a simplified model. There are at least the following four important real-life complications:

•Most importantly, Figure 6 neglects the time to detect Soviet missile launch, transmit this information up the chain of command, make a command decision to dash, transmit this decision to the HMLs, start the HMLs, deploy them in hardened configurations at the end of their dash, and (hopefully) give the drivers time to escape. Thus the actual time, from Soviet missile launch to HML deployment, would need to be at least several minutes longer, and perhaps a lot longer, than the times shown in Figure 6 in order to attain a given dispersal area. In fact, some experts believe there will be no time for any dash at all, that "In the event of a nuclear attack, the first information the military receives is likely to be the last clear information they'll get; we will get that clear information about what is going to happen to us about the time the nuclear warheads hit" (22).

•Most of the dash would not proceed directly inward or outward but would instead be along available roads with a last-minute move off roads. This effect of course increases the dash time needed to attain a given dispersal area. It could be very significant if there were only a few roads leading from the peacetime to the alert and attack areas, and less significant if the Air Force built very many roads. Calculations with an idealized but plausible road net-work (8 roads leading radially outward from each alert area, with 16 connecting roads between them) show that, in order to expand the alert area outward by 50%, a 30-minute dash is needed rather than the 10-minute dash found from Figure 6. So this effect is likely to be large.

•The HMLs might make no attempt to dash from their initial peacetime or alert positions, because they would probably be initially parked, perhaps in a hardened configuration, in which case it might be safest to just leave them there (11).

•The alert and peacetime regions are not circular. Calculations show that the dash time is not very sensitive to the shapes of the alert regions, so long as the alert regions are not highly elongated. An elongated region leads to shorter dash times (to attain a given dispersal area) because the dash starts from a longer perimeter. For example, the dash time needed to expand the deployment area outward by 50% would be cut in half (from 10 to 5 minutes) if all five alert areas were rectangles having (length) = 4 X (width).

The first two complications will certainly be present; their net effect is to increase the required dash time (to attain a desired dispersal area) by at least several minutes, and perhaps by 30 minutes or more. The last two complications might or might not make a difference. The only one that reduces the required dash time is the last of the four. It is fair to conclude that the real time, starting from Soviet missile launch, needed to reach a given deployment area, would be at least several minutes longer, and perhaps very much longer, than the times shown in Figure 6.

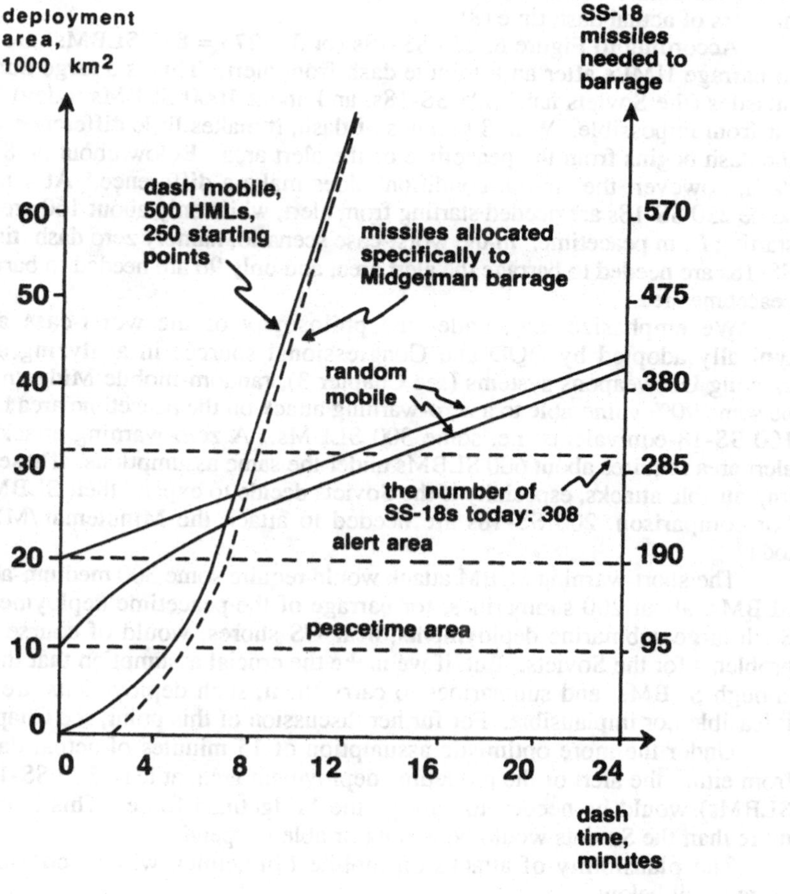

Figure 6. Deployment area generated, and number of SS-18-equivalents needed to barrage that area at an 85% HML destruction rate, as a function of dash time, for the dash-mobile and random-mobile basing modes. The attacking warheads are assumed to be air-burst at optimum height. The assumed dash speed is 45 km/hr (28 mph). "Dash time" means the time actually spent moving at this average speed. The total time to deploy, including warning, communication, start-up, and deployment in hardened configuration, would be longer. For dash-mobile, the dashed line gives the number of "excess" missiles needed, above those needed to target the Minuteman silos at which the HMLs are based.

These times should be compared with the 15 minute SLBM flight time to the central US from 200 miles offshore, or 8 minutes with depressed trajectories. Clearly, the actual time available for the dash (after detecting the launch, receiving the "go" signal, etc.) could be only 4-8 minutes, or even zero. A Congressional Research Services report states that there might be only 3-6 minutes of actual dash time (8).

According to Figure 6, 275 SS-18s (or 3 x 275 = 825 SLBMs) are needed to barrage HMLs after an 8-minute dash from alert. This is a large number of missiles (the Soviets have 308 SS-18s, and about 1000 SLBMs today) but it is far from impossible. With 8 minutes of dash, it makes little difference whether the dash begins from the peacetime or the alert area. Below about an 8 minute dash, however, the "initial condition" does make a difference. At 4 minutes, some 250 SS-18s are needed starting from alert, while only about 160 are needed starting from peacetime. In the worst-case scenario, namely zero dash time, 192 SS-18s are needed to barrage the alert area, and only 96 are needed to barrage the peacetime area.

We emphasize that, under the philosophy of the worst-case analysis typically adopted by DOD and Congressional sources in analyzing already-existing US weapons systems (see Chapter 3), random-mobile Midgetmen will be some 90% vulnerable to a zero-warning attack on the peacetime area by about 100 SS-18-equivalents, i.e. some 300 SLBMs. A zero-warning attack on the alert area requires about 600 SLBMs under the same assumptions. These are not implausible attacks, especially if the Soviets decide to expand their SLBM force. For comparison, 200 SS-18s are needed to attack the Minuteman/MX force today.

The short-warning SLBM attack would require some 300 medium-accuracy SLBMs, about 200 submarines, for barrage of the peacetime deployment area. Such large submarine deployments, near US shores, would of course present problems for the Soviets. But, if we make the crucial assumption that they have enough SLBMs, and submarines to carry them, such deployments are neither infeasible nor implausible. For further discussion of this point, see Chapter 3.

Under the more optimistic assumption of 15 minutes of actual dash time from either the alert or the peacetime deployment area, at least 300 SS-18s (900 SLBMs) would be needed to barrage the Midgetman force. This is probably more than the Soviets would be willing or able to spend.

The plausibility of attacks on mobile Midgetmen will be considered in more detail below.

Maneuvering directed warheads

DOD has worked for several years on early studies of methods of destroying "relocatable targets," such as the rail-mobile SS-24 and the land-mobile SS-25 (similar to Midgetman). One such concept is an ICBM re-entry vehicle that would receive surveillance satellite signals and then maneuver in the atmosphere, perhaps aided by a terminal homing system (23-27). Such maneuvering re-entry vehicles, regardless of how they are directed, are called "MaRVs." The primary destruction mechanism might be microwave radiation from a nuclear explosion, rather than blast (28). A second concept is to approach the target with Stealth bombers, perhaps with assistance from surveillance satellites, and then destroy the targets with short-range attack missiles or with cruise missiles that would seek out and home in on moving targets (29-33). Of course, similar devices could be developed by the Soviets for use against Midgetman or rail-based MX.

A negotiated ban on the flight-testing of MaRVs would provide some insurance against the future vulnerability of mobile missiles. Such a ban would preclude Soviet deployment of the first concept mentioned above, but not the second.

The use of the B-2 Stealth bomber in a nuclear war-fighting role to destroy mobile missiles has received much attention (31,32) and criticism (33) lately. B-2 deployment is currently scheduled to begin in 1992. It is far from clear that it will be able to destroy mobile missiles, because the surveillance satellites on which it depends for guidance might be destroyed early in a war, while use of its own electronic sensors for guidance would give away the bomber's position. If Stealth technology does turn out to be a feasible way to locate and destroy mobile missiles, it will pose a threat to our own mobiles once the Soviets develop it.

Dash-mobile vulnerability

If deployed, the earliest Midgetman basing mode would probably be HMLs based at Minuteman silos at Malmstrom Air Force Base, Montana (11). The current plan is to locate 150-250 HMLs in soft bunkers among the 200 Minuteman silo sites at Malmstrom, with two HMLS at each silo site (i.e. there would be 75-125 HML sites) (Figure 7). "In the event of imminent attack" (5), the HMLs would disperse onto the road system that links and surrounds Malmstrom's Minuteman system. The system is designed to be able to disperse a "safe distance" from the Minuteman silos "within the flight time of an attacking reentry vehicle," i.e. presumably within a 30-minute ICBM flight time. According to Air Force officials, when this system is installed it will be the only US land-based, strategic missile designed to survive an attack without having to launch on warning (11).

A targeted attack could be attempted against this basing mode, hoping to catch HMLs in their bunkers. This attack would be "free" for the Soviets because no extra warheads would be needed beyond those already needed to destroy US Minuteman silos. However, the Soviets would, no doubt, devote extra warheads to barraging as large an area as feasible surrounding each of the HML sites, in case the HMLs got out of their bunkers. Thus this attack can be analyzed in the same way as the general area barrage analyzed above. The only difference is that we now have many small area barrages instead of five larger ones. For consistency with our analysis of random-mobile, we assume that 500 HMLs are deployed in the dash mobile mode, so that there are 250 bunkers and thus 250 circular areas to be barraged.

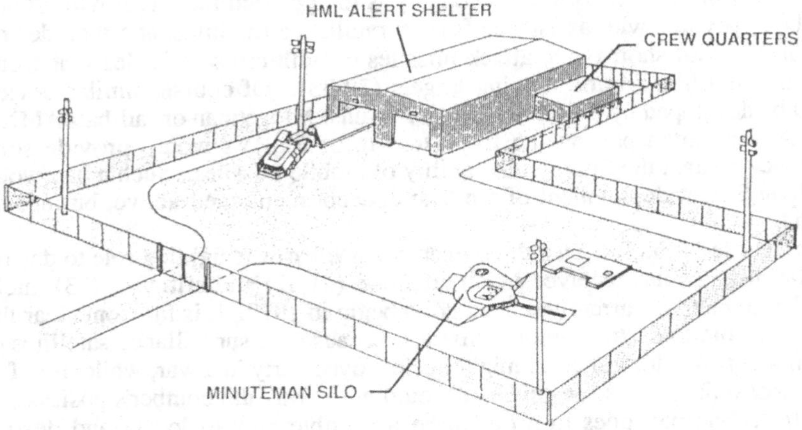

Figure 7. Present plan for initial deployment of the Midgetman missile calls for collocating the missiles within a secure area of existing Minuteman silos. Crews for the hard mobile launchers would have quarters next to the bunkers so that they could deploy the missiles onto road networks if an attack was launched. (Drawing courtesy of US Department of Defense.)

We begin from the fact that 9.5 SS-18-equivalents per 1000 km^2 are needed to barrage the 500 HMLs at some 85% effectiveness; during dash time t each HML disperses over a circle of area $A(t) = \pi v^2 t^2$ surrounding its bunker, at a dispersal speed v=47 km/hr; there are 250 such circles to be barraged.

Thus we calculate the dash-mobile graph (solid line) of Figure 6. The graph shows the total number of missiles needed to barrage the area generated after dash time t. But some of this barrage would be supplied "for free" by the warheads needed to attack the 250 Minuteman (MM) silos in an attack on the full MM/MX/Midgetman system. This targeted attack on silos contributes on the order of 250 0.5-MT warheads, or 25 SS-18-equivalents, to the barrage. So the number of missiles devoted specifically to the Midgetman barrage is 25 less than the solid-line graph. The dashed-line graph.shows the number of warheads specifically allocated to the Midgetman system, over and above those used against the Minuteman system.

It is worth noting that, during the first 2 minutes of actual dash, no excess warheads are needed since the Midgetman will still be within destruction distance of the Minuteman silos. HMLs dash at about 0.75 km/min, and the radius of destruction of a 0.5 MT warhead (the SS-18 warhead yield) against a 2 atm (30 psi) HML is calculated from Chapter 3 Eq. (5) to be about 1.5 km.

Once again, the graph is based on simplified assumptions. In real life:

•Time will be lost in receiving the dash command, getting started, and deploying.

•The dash would proceed mainly along available roads.

•If the selected bunker sites are near each other, the circular areas will begin to overlap after a short time so that the total dispersal area will actually be less than estimated in Figure 6. The average spacing between Minuteman silos is roughly 8 km, so this effect would begin showing up after 6 minutes of dash.

So Figure 6 is optimistic.

Figure 6 highlights the critical difference between the two mobile modes: Dash-mobile can be surprised by a very small barrage at short times. As we have noted above, if surprised with under 2 minutes of dash this mode can be destroyed "for free," with no excess missiles beyond those needed anyway to attack the MM silos. With six minutes of dash, a plausible time under preemptive SLBM attack (8), 95 "excess" missiles are needed, whereas over 200 would be needed to attack random-mobile. On the other hand, above some 10 minutes of dash this mode is dispersed over a much larger area than is the random mobile mode.

Briefly, in the worst case the dash mobile mode will be vulnerable to a short-warning attack that costs the Soviets nothing beyond those warheads needed to attack the MM/MX system. In the best case (12 or more minutes of actual dash), an implausibly large number of SS-18-equivalents, over 600, would be needed to barrage this mode, i.e. the HMLs would be safe.

Mobile missiles will not be synergistically survivable with bombers

Although the MM/MX silo-based force is today more-or-less vulnerable to Soviet ICBMs, it is not vulnerable to SLBMs. Thus US bombers and ICBMs are today synergistically survivable: they cannot be simultaneously destroyed, even though they are separately vulnerable. But as we saw in Chapter 3, more accurate Soviet SLBMs might spell the end of synergistic survivability by the mid-1990s.

Mobile Midgetman will not rectify this, although it will increase the number of SLBMs needed for the simultaneous attack. If Midgetman is deployed in mobile launchers, the US might be faced in the 1990s with a situation in which a large Soviet SLBM force, perhaps employing depressed trajectories, could threathen US bombers and HMLs with a short-warning barrage, and simultaneously threaten the MM/MX force with a short-warning targeted attack. If they chose to pursue such a strategy, the Soviets would need to deploy a highly accurate SLBM to attack the MM/MX force, to deploy greater numbers of SLBMs, and to routinely station large numbers of strategic submarines near US shores. For further discussion of this point, see Chapter 3.

As we will see in Chapter 11, superhard silos would rectify the synergistic survivability problem. During this century, superhard silos will be vulnerable only to an attack by heavy ICBMs carrying high yields at very high accuracies. SLBMs will not be able to destroy superhard silos. Thus Midgetman in superhard silos will be synergistically survivable with bombers, while mobile Midgetman will not.

Future Soviet nuclear weapons developments could be influenced by US Midgetman decisions: A decision for mobile Midgetman deployment could stimulate more Soviet SLBMs and depressed trajectories. A decision for superhard silo deployment would encourage continued Soviet reliance on ICBMs, and the development of large and very accurate ICBM warheads.

Survivability depends on arms control

It is probably true that every land-basing scheme is vulnerable. Midgetman was never conceived as an absolutely invulnerable force, in the sense that US ICBMs were once absolutely invulnerable. One hopes, at best, for a deployment scheme that makes Soviet attacks implausible. So Midgetman survivability depends on the plausibility of various attack scenarios, which depends on the numbers and kinds of Soviet missiles, which in turn depend on US posture and on the arms control situation.

Thus we study ICBM survivability under four arms control scenarios: an *unconstrained arms race*, *SALT-II limits*, *50% cuts* below SALT-II limits, and more extreme *finite deterrence* cuts. We assume that the US ICBM force includes 500 mobile Midgetmen and, except for the finite deterrence case, 1000 silo-based Minutemen and MXs.

An *unconstrained arms race*, of course, makes Midgetman survival prospects dismal. Clearly the Soviets could build enough missiles to destroy not only the entire MM/MX force but also a mobile Midgetman force deployed over any plausible area.

Specifically, suppose the Soviets doubled their present force (see Table 1 of Chapter 1) to 2800 ICBMs and 2000 SLBMs. Without arms control limits, such a doubling is quite plausible by the mid-90s, especially if the Soviets refurbish their retired missiles (8). Assuming a plausible distribution of the various types of Soviet missiles, this ICBM force would have a barrage effectiveness (Table 2) of some 1800 SS-18s, and the SLBM force would be equivalent to 700 SS-18s. The Soviets could choose to skew this distribution toward more SLBMs, in response to US mobile deployment.

This Soviet force is more then enough to destroy the entire US MM/MX/Midgetman force. As we have noted, the most plausible attack is from submarines. 2000 high-accuracy SLBM warheads (250 SLBMs) could destroy the 1000 MM/MX silos, while 900 SLBMs (300 SS-18-equivalents) barraged the Midgetman force with 8-12 minutes of dash (see Figure 6). The remaining 800 SLBMs could be held in reserve or used against bomber bases. In this scenario, the main benefit of Midgetman is to raise the price to attack US ICBMs

Mobile Midgtman fares especially poorly under no arms constraints, more poorly for example than superhard silos, for two reasons: Increases in Soviet force size make large barrage attacks more plausible, whereas superhard silo vulnerability is not so dependent on Soviet force size, being dependent mainly on the lethality of individual Soviet warheads. Secondly, Soviet increases cannot be countered by Midgetman increases, because additional Midgetmen will still be restricted to the same deployment area due to lack of suitable DOD land, and it is the size of the deployment area, not the number of Midgtmen on that area, that primarily determines survival.

The United States should not plan to rely on land-mobile missiles unless it plans also to prevent an unconstrained arms race.

Under *SALT-II limits*, the Soviets might have something like their present 1400 ICBMs and 1000 SLBMs, perhaps skewed more toward SLBMs, by the mid-90s. The present ICBM force has a barrage effectiveness of some 900 SS-18s, and the SLBM force has an effectiveness of some 350 SS-18s.

In a worst-case scenario in which dash-mobile Midgetmen are surprised with less than two minutes dash from their MM enclosures, Soviet submarines could destroy the full US ICBM force using only 350 of their 1000 SLBMs in a high-accuracy 2-on-1 attack against MM/MX silos that also destroys the Midgetmen.

Even allowing the dash mobile mode 12 minutes of warning, the Soviets could devote 600-900 SLBMs to destroying the MM/MX/Midgetman force. This attack would trade most Soviet SLBMs for most US ICBMs, leaving a few hundred Soviet SLBMs for an attack on bomber bases, and leaving all Soviet ICBMs in reserve. For the Soviets, it would be a good trade in terms of missiles (600-900 SLBMs for some 1500 US ICBMs) but a bad trade in terms of warheads (some 5000 SLBM warheads for some 2000 US ICBM warheads). This warhead-exchange ratio might deter Soviet preemption.

Random-mobile Midgetman fares much better. In the worst case, 300 Soviet SLBMs might surprise the HMLs on their peacetime operating area, while another 350 high-accuracy SLBMs might destroy the MM/MX force. This would leave several hundred SLBMs for attacks on bomber bases. However, if the Midgetmen were first put on alert, 600 Soviet SLBMs would be needed to barrage them, and still more would be needed if the Soviets wanted to allow for some HML dash time. The Soviets might then have enough SLBMs left to attack the MM/MX silos (350 high-accuracy SLBMs), but they would have to expand their submarine force beyond 1000 SLBMs if they wanted to have enough to also destroy US bombers. That is, if random-mobile HMLs are caught near their alert area, a Soviet attack could trade the Soviet SLBM force for the US ICBM force. 5 or 10 minutes of dash from alert status, however, would preserve a significant fraction (perhaps 50%) of the HMLs.

Three points should be emphasized: First, random-mobile does not have the extreme worst-case instability noted for dash-mobile. Second, mobile Midgetmen in a SALT-II-limited environment might encourage the Soviets to further favor SLBMs at the expense of their ICBMs, a trend that is evident today (34). Third, mobile missile survival under SALT-II limitations depends strongly on adequate warning: HMLs must have both strategic warning, to go to alert status, and then sufficient tactical warning for a dash of some 10 minutes from the alert area onto surrounding land. The President might hesitate to order either of these moves, and if they are ordered they might prove de-stabilizing in a crisis.

With *50% cuts*, a Soviet SLBM attack against the full US ICBM force begins to look very implausible. As an approximation of 50% cuts, assume that each side is allowed 1200 ballistic missiles, with a maximum of 900 in either the ICBM or the SLBM leg. The required Soviet cuts would probably come more from their ICBMs than their SLBMs, due to greater SLBM survivability, especially if they develop a highly accurate SLBM. We thus assume that the Soviets retain 600 ICBMs (including 150 SS-18s) and 600

SLBMs. The ICBMs would have a barrage effectiveness of some 400 SS-18-equivalents, while the SLBMs would have an effectiveness of 200 SS-18s.

The dash-mobile mode would then still exhibit its worst-case instability: 350 SLBMs could attack the MM/MX silos in a 2-on-1 attack hoping to surprise the HMLs with under 2 minutes of dash from their Minuteman enclosures. Using all 600 of their SLBMs, the Soviets could destroy the MM/MX force and the HMLs with under 6 minutes of dash. With more than some 6 minutes of dash, some of the HMLs would escape.

600 Soviet SLBMs would be enough to barrage the peacetime random-mobile area with enough left over for a 2-on-1 attack on MM/MX silos. However, the alert area could only be barraged by the full 600-missile force, leaving no SLBMs to attack MM/MX silos. That is, the Soviets could trade their SLBM force for our ICBM force if , but only if, they surprised the HMLs on their peacetime area.

Thus, with 50% cuts and random-mobile Midgetman deployment, the massive SLBM attack needed to destroy all US ICBMs looks implausible, even in a crisis. But the dash-mobile deployment remains unstable even under 50% cuts.

An ICBM attack on all US ICBMs and bombers would be theoretically possible, but would give the Midgetmen many minutes warning due to the 30-minute ICBM flight time. A new form of synergistic survivability would exist: The Soviets might destroy the MM/MX force and the bombers by using most of their SLBMs in a short-warning attack, but they would not have enough SLBMs to also barrage Midgetman. If they used ICBMs for this barrage, the Midgetmen could escape prior to ICBM arrival. Furthermore, Midgetman missiles could be fired prior to ICBM arrival but following Soviet SLBM explosions.

It has often been noted (35) that some few hundreds of survivable warheads on each side would preserve superpower nuclear deterrence. Thus, let us assume a *finite deterrence* force of 2000 warheads on each side (36). A plausible deployment might be 500 SLBM warheads and 1000 bomber warheads on each side. Assuming 0.5 MT warheads, the entire Soviet SLBM force would be needed to target the 500 dash-mobile Midgetman locations, while the full Soviet missile force, SLBMs and ICBMs, would be needed to barrage just the peacetime random-mobile operating area. These attacks, which depend on complete surprise and which allow all US bombers and SLBMs to survive, are quite implausible. On the other hand, 500 sufficiently large Soviet warheads, either on SLBMs or ICBMs, could barrage random-mobile Midgetmen. For example, just 300 10-MT warheads are enough to barrage the random-mobile alert area. A limit of, say, 1 MT on warhead yields would solve this problem: 500 1-MT warheads could barrage only 70% of the peacetime operating area.

Thus finite deterrence is by far the safest scenario for US ICBM survivability. Coupled with a one megaton yield limit, random-mobile Midgetman would be essentially invulnerable on the peacetime deployment area. In a crisis, the US would under no pressure to issue possibly destabilizing orders to dash onto a large area, and no pressure to launch before Soviet missiles began exploding on US soil.

Summary, discussion, conclusions

Because they depend on adequate tactical warning, mobile Midgetmen could be attacked by off-shore submarines rather than by ICBMs. This makes them subject to the same short-warning SLBM attack that would be used against US bombers.

Mobile missile vulnerability to SLBMs might be of critical importance in the 1990s when the Soviets may have a high-accuracy, silo-destroying SLBM similar to the US Trident II. Such a development would make US bombers and silo-based ICBMs vulnerable to a simultaneous SLBM attack, in contrast to today's synergistic survivability of these two forces. Mobile Midgetmen, vulnerable to the same short-warning SLBM attack, may do little to rectify this problem.

Maneuvering re-entry vehicles guided toward individual mobile launchers could make Midgetman highly vulnerable. Although the Soviets are not likely to develop such "MaRVs" during this century, an arms control agreement banning MaRV flight tests would provide insurance against this development, and would extend Midgetman's deterrent value further into the next century.

Deployment at Minuteman silos is much less stable than deployment over large tracts of DOD land. If surprised with only a few minutes of dash, the dash-mobile mode is vulnerable to barrage attacks using only some 50% more missiles than are needed anyway to attack the Minuteman silos. Under SLBM attack, such short tactical warnings are plausible.

Current plans call for initial deployment of 250 Midgetmen in the dash-mobile, not random-mobile, mode (11).

Random-mobile deployment is more stable. If surprised with no strategic warning and no tactical warning it is vulnerable to a barrage by as few as 100 SS-18-equivalents (300 SLBMs), but with either a 5-minute dash from the peacetime area, or strategic warning allowing a shift to the alert area, 200 SS-18-equivalents (600 SLBMs) are needed. For comparison, 200 SS-18s is the number number today to attack the silo-based MM/MX force.

Thus, US deployment of 500 Midgetmen in the random-mobile mode would roughly double the price to attack the entire US ICBM force, from 200 SS-18s to 400 SS-18-equivalents. It furthermore shifts the attacking force from ICBMs to SLBMs, and requires that the Soviets mass a large number of submarines within a few hundred miles of US shores.

Midgetman survivability depends strongly on adequate warning, and on the President ordering alert status and , upon receiving word of Soviet attack, orders to dash. This dependence on warning and alert deployment is de-stabilizing: Signs of alerted or dashing launchers could prompt the Soviets to pre-empt; the President might hesitate to issue such orders, leaving the launchers exposed to a small (and thus tempting) Soviet barrage; Soviet awareness of US plans for alert and dash status could shape Soviet plans in the direction of off-shore SLBM deployment and pre-emption at an earlier stage in a developing crisis. Briefly, Midgetman's dependence on warning and on alert deployment encourages surprise-oriented strategies.

Survivability of mobile Midgetman is strongly linked with arms control. Even with 500 mobile Midgetmen, the full MM/MX/Midgetman land-based force will still be vulnerable to an attack of roughly twice the size that land-based missiles are vulnerable to today. If the Soviets have enough missiles of the right type, such an attack might be attempted in a crisis. With no arms control, the Soviets could obviously have enough missiles. Even under today's SALT-II limits, they have enough SLBMs to mount this attack. But under 50% cuts, the Soviets would have only a very marginal ability to carry out such a large attack, so they would probably not attempt it in the first place. Under a 2000-warhead "finite deterrence" arms control regime, 500 random-mobile Midgetmen would be absolutely invulnerable. Furthermore, finite deterrence would allow the US to dispense with destabilizing alert and dash plans.

Mobile Midgetman makes sense if, but only if, coupled with arms control cuts on the order of 50% or more.

References and notes

1. US Air Force Scientific Advisory Board, *Report of the Small Missile Independent Advisory Group*, USAF, Washington, D.C. Sept 1983.

2. US Air Force , *Fact Sheet: Small Intercontinental Ballistic Missile*, USAF, Washington DC, June 1985.

3. US Air Force news release, "Air Force surveys more basing modes," USAF, Nov 1985.

4. *ICBM Modernization Program*, Annual Progress Report to the Committees on Armed Services, Washington DC, 15 Jan 1985.Jan 1985.

5. General Accounting Office, *Status of the Intercontinental Ballistic Missile Program*, US GAO, 8 July 1985.

6. US General Accounting Office, *ICBM Modernization*, GAO, Washington, DC. Sept 1986.

7. Jonathan Medalia, *Small Single-Warhead ICBMs*, Congressional Research Service Report, 26 May 1983.

8. Jonathan Medalia, *Midgetman Small ICBM: Issues Facing Congress in 1986*, Congressional Research Service Report, 20 March 1987.

9. Jonathan Medalia, *MX, Midgetman, and Titan Missile Programs*, Congressional Research Service Issue Brief, 26 March 1987.

10. Aviation Week and Space Technology is a good source of technical information about Midgtman and its HML.See, e.g., 18 May 1987, pp. 47-48; 5 Jan 1987, pp. 20-21; 9 Mar 1987, pp. 31-37.

11. Aviation Week and Space Technology, "USAF plans single launch control facility for first small ICBM force," 18 May 1987, 47-48.

12. R. Jeffrey Smith, "A scheme to attract missiles and deter an attack," Science 27 June 1986, p. 1592.

13. Matthew Bunn, *Technology of Ballistic Missile Reentry Vehicles*, Program in Science and Technology for International Security, MIT, Cambridge, March 1984.

14. H. Lucas, Jane's Defense Weekly, 19 Jan 1985, p. 96.

15. Thomas B. Cochran, William M. Arkin, Milton M. Hoenig, *Nuclear Weapons Databook Vol. I: US Nuclear Forces and Capabilities*, Ballinger Pub. Co., Cambridge, Mass. (1984).

16. Kosta Tsipis, *Arsenal*, Simon and Schuster, New York (1983), pp. 277-278.

17. Samuel Glasstone and Philip Dolan, *The Effects of Nuclear Weapons*, US Departments of Defense and Energy, Washington, DC (1977).

18. I thank Herbert Nelson of the Naval Research Laboratory for these calculations.

19. Ian Bellany, *Nuclear Vulnerability Handbook*, Center for the Study of Arms Control, Lancaster, England (1981).

20. I thank John Michener for pointing this out.

21. Barton Wright, *Soviet Missiles: Data from 100 Unclassified Sources*, Lexington Books, Lexington (1986).

22. Interview with Ashton Carter, "Are we prepared to fight a nuclear war?" USA Today Magazine, Dec 1983, p.6.

23. Aviation Week and Space Technology, "USAF studies mobile-target ballistic missile," 23 July 1984, p. 18.

24. Bill Keller, "US missile plans aims to penetrate any defenses Moscow can devise," International Herald Tribune, 12 Feb 1985, p.1.

25. "Ways to attack mobile missiles under study," Defense Daily, 1 Apr 1985, p. 171.

26. Ashton Carter, "Satellites and anti-satellites: the limits of the possible," International Security, Spring 1986, p. 69.

27. R. Jeffrey Smith, "Proposal to ban missiles favors targeting over arms control," Science, 22 Aug 1986, pp. 831-833.

28. William M. Arkin, "Test ban fever," Bulletin of the Atomic Scientists, Oct. 1986, pp. 4-5.

29. Aviation Week and Space Technology, "New cruise Missile study," 9 Feb 1987, p. 24.

30. William M. Arkin, "The new mix of defense and deterrence," Bulletin of the Atomic Scientists, June/ July 1986, pp. 4-5.

31. Aviation Week and Space Technology, "USAF, Northrop unveil B-2 next-generation bomber," and succeeding articles, 28 Nov 1988, pp. 20-27.

32. Aviation Week and Space Technology, "B-2 bomber development," 5 Dec 1988, pp. 18-22.

33. For discussion and critique of the B-2's role and capabilities, see John Pike and David Bourns "The Stealth bomber--its rationale is also invisible," Federation of American Scientists Public Interest Report, Oct 1988, pp. 1-12.

34. Robert Gates, Lawrence Gershwin, *Soviet Strategic Force Developments*, Senate Armed Services Committee, 26 June 1985.

35. Robert McNamara, *The Fiscal Year 1969-73 Defense Program and the 1969 Defense Budget*, Department of Defense, Washington DC (1968).

36. Harold A. Feiveson, Richard H. Ullman, Frank von Hippel, "Reducing US and Soviet nuclear arsenals," Bulletin of the Atomic Scientists, Aug 1985, pp. 144-150.

Part V Chapter 8

Rail garrison MX

Peter D. Zimmerman

Introduction

The principal reasons for deploying the MX missile in a rail garrison mode are to hold a larger number of hardened Soviet targets at risk with only a small delay and to reduce the vulnerability of silo-based missiles to attack. A further reason for deploying a new ICBM is, of course, the age of the Minuteman III (MM III), the principal US land-based strategic missile. Intended to have a 10 year service life, the MM III is now approaching 30 years of service.

The survivability of any target subjected to nuclear attack is influenced by three characteristics, which may be thought of as forming the axes of an orthogonal co-ordinate system, properly scaled. These characteristics are as follows:

•The hardness of the target; how readily it can be destroyed by blast, heat and nuclear effects.

Figure 1. Artist's concept of a typical rail garrison layout. [Official US Air Force drawing.]

•The speed of the target; how rapidly it can change location so that it can escape the destruction region of the attacking missile.

•The uncertainty in location of the target; its ability to hide so that the enemy cannot attack without expending a disproportionate number of warheads.

One possible figure of merit, the magnitude of a vector in the "survivability space," for evaluating the survivability of a strategic weapon is the number of attacking warheads needed to have a single-shot kill probability, SSKP (= 1-SSPS; see Chapter 3), greater than some set value, say 0.9, against one retaliatory delivery vehicle or, conversely, the kill probability of a single warhead fired at a second-strike weapon. For US ballistic missile submarines operating in the broad ocean areas of the world (i.e. submarines carrying Trident missiles), SSKP is practically zero. Choosing the number of weapons needed to have at least a SSKP of 0.9 to kill a strategic nuclear delivery vehicle as defined in the SALT II Treaty as our figure of merit, we obtain (roughly) the values given in Table 1.

An alternative criterion might have been the number of warheads required to destroy a single warhead. This is less satisfactory because the targets of an attack are the nuclear delivery vehicles. However, choosing the warhead-to-warhead ratio emphasizes the fact that single-warhead missiles are less attractive targets than MIRVed missiles. With such a choice, the figure of merit for silo-based ICBMs would range from 1 to 2/10 -- the values obtained from a single shot against a one-warhead ICBM and two warheads used against a silo containing a missile with ten warheads. The figures of merit for MIRVed SLBMs and rail-mobile missiles in garrison would also be reduced by their fractionation value.

The present strategic triad consisting of fixed ICBMs, bombers which can dash on warning, and submarine-launched ballistic missiles lie, in effect, along principal axes in this survivability space. The ICBMs can be enormously blast-resistant (up to 80,000-100,000 psi) if placed in superhardened silos, but they are at fixed well-known locations. Bomber aircraft are relatively soft, take off from well-known airfields, but are able to move fast enough to be untargetable after

Figure 2. A typical rail garrison configuration houses up to four missile trains on railroad tracks in side-by-side protected shelters. The garrisons are located on selected Air Force bases. [US Air Force drawing.]

Table 1. Vulnerability space for strategic deployments

System	Figure of merit
Silo-based ICBM	1-2
SLBM (present anti-submarine capabilities)	infinite
SLBM (submarine localization possible)	10-20
Bombers (on alert, under SLBM attack)	small number[a]
Bombers (after takeoff)	infinite
Dash mobile Midgetman (parked at Minuteman silos)	2/3[b]
Dash mobile Midgetman (dash under attack)	5[c]
Dash mobile Midgetman (90 minutes dash)	infinite
Rail mobile MX (in garrison)	1/4[d]
Rail mobile MX (in garrison, 90 min ICT[e])	30[e]
Rail mobile MX (sent out on strategic warning)	>100

Notes for Table 1.

a. If a group of 10 bombers is awaiting take-off, some will probably be caught on the ground or within a short distance of the airfield by an air-burst. A 0.4 MT airburst produces free air overpressures of 5 psi (sufficient to damage an aircraft) at distances of 3.5 miles. A bomber shortly after takeoff might fly at a speed of about 400 mph, so that it would cover 3.5 miles in about half a minute. The fly-out routes for such aircraft are, however, relatively predictable, so a barrage might be directed against relatively linear paths rather than be required to demolish an entire area. The number of warheads required to destroy aircraft on the ground is clearly small, so the survivability of the bomber component of the Triad depends sensitively upon having good tactical warning and excellent C^3 (command, control, and communications).

b. Includes the bonus of getting one silo-based missile and two HMLs with the same two-on-one attack.

c. Assumes 5 minutes dash time for each of two HMLs based at a single Minuteman silo and that area is generated in a simple way. HML hardness: 35 psi; weapon yield 400 kt.

d. Assumes four trains, each with two missiles, can be destroyed by a single weapon but that the garrison must be attacked by at least two warheads to account for possible unreliability.

e. Assumes that the intelligence cycle time (ICT) plus the missile flight time for a Soviet attack is 90 minutes, considered a reasonable minimum for satellite observation. The Soviets must, under this circumstance, barrage at least the first 90 minutes of track which can be generated since they cannot know if the trains remain in garrison. It is assumed that both missiles on a train are destroyed.

the first few minutes of flight. Finally, submarines are relatively soft (particularly when cruising at great depth), cannot move rapidly (particularly if they wish to remain silent), but have almost perfect concealment after they leave port (1).

Most schemes for ensuring the survivability of strategic weapons have tended to lie along the principal axes of the survivability space, in effect seeking the same kinds of protection as the present Triad. It is worth considering deployment methods which are linear combinations of the three principal vectors. The rail-garrison MX is one such. It obtains its survivability through moderate hardness (ranging from 5 to 90 psi), moderate dash speed (50-100 km/hr) and good but not perfect concealment once deployed (2).

Vulnerability criteria

Rail vehicles have a range of strengths depending, primarily, upon their orientation relative to the blast wave and secondarily upon their construction and the type of vehicle (locomotive, box car, etc.). It is not surprising that trains are significantly harder when the blast strikes the locomotive head on than when the blast wave is incident perpendicular to the rails. In the first case there is little torque to derail the train; in the second case the entire force of the shock wave acts to overturn the car.

Experiments with above-ground explosions at the Nevada test site as well as surveys conducted after the Hiroshima and Nagasaki attacks provide reasonable values to be used in calculating the survivability of rail targets. Glasstone and Dolan published graphic photographs of loaded wooden boxcars subjected to 4 and 6 psi overpressures in calibrated tests in Nevada (3). The car subjected to only 4 psi is substantially undamaged and able to roll, while the one subjected to 6 psi has had large portions of its siding removed. Importantly, however, the more severely damaged car is still upright with all of its wheels on the track. With only minor work to remove a dragging panel, it could be moved. A similar car was overturned at an overpressure of 7.5 psi, and a fourth car was demolished at an overpressure of 9 psi. There was no damage to the track. Transportation equipment is generally classed as drag-sensitive targets (4).

A diesel locomotive weighing 46 tons with its engine running was exposed to 6 psi overpressure from a nuclear blast at the Nevada test site; it continued to operate normally afterwards (5). More generally, the vulnerabilities of rail equipment to blast can be estimated from Figure 5.146 of Effects and the accompanying table on page 222. For a constant weapon yield of 0.5 MT, Table 2 gives survivability distances and approximate strengths (assuming optimum height of burst).

Table 2. Radius of destruction of a 0.5 megaton warhead against railroad targets.

Vehicle	Range (km)	Strength (psi)
Boxcar, side on	5.5	5-6 psi
Locomotive, side on	3.1	12-15 psi
Boxcar, end on	1.9	20-25 psi
Locomotive, end on	1.3	40-50 psi

More modern diesel locomotives are expected to have end-on strengths up to 90 psi (6), although reasonable strategic planners must assume the lower values characteristic of side-on exposure. Aircraft, as an aside, seem to be only 2.4 to 3.6 psi hard (7).

Air Force figures for the survivability of the rail-garrison vehicle assume that a properly-placed warhead can provide a blast overpressure of 5 psi over a 7 km length of track (see Part IV Chapter 6). For once Air Force damage criteria are probably overly pessimistic, because non-perpendicular shock waves are significantly less capable of damaging the vehicles, and because modern steel railcars specially designed for service in a missile train could be somewhat harder than 5 psi. We will, therefore, take as the damage criterion for trains out of garrison a 5 psi component of the shock wave in the direction perpendicular to the track.

Weapons effects

The radius from ground zero at which one finds a given blast overpressure varies as $Y^{1/3}$, but blast overpressure does not fall off as R^{-3} as is often erroneously stated (8). Indeed, the relationship is far more complex and not readily given by a simple equation. This is because of the complex interaction of the free-field shock wave, the shock wave reflected from the ground, and the heating of the air because of compression and the heating of the ground because of the direct radiation from the fireball (9). The problem becomes especially complex for varying heights of burst (HOB); in general there is a given HOB which will maximize the area subjected to at least a given overpressure for a weapon of specified yield. Caution is advised in using the circular slide rule packed with Effects; the overpressure/range relationship is correct only for continuously varying HOB, the HOB chosen at all points to be optimum for the given range and overpressure.

The approach taken in this paper was to choose a yield considered plausible for an SS-18 MIRV warhead (400 kt) (10), to optimize the HOB for roughly 30

Figure 3. Close-up view of two MX train cars inside a rail garrison, with two combat crewmen at their work stations in the missile launch control cars. [US Air Force drawing.]

psi overpressure, and then to fit detailed calculations of the overpressure at varying ranges from ground zero to an analytic form. The form chosen was:

$$P(r,Y) = Y^{1/3}[A/r^3 + B/r^{3/2}] \qquad (1)$$

where the overpressure P is in psi, yield Y is in kilotons, range r is in kilometers. The value of A is 2.98 and that of B is 4.61. This equation reproduces more accurately computed results to within 0.5 psi over a range from 12 psi to 3.2 psi and over radii from 2.25 to 5 km. It is better in the far field than in the near field. The choice of a 30 psi overpressure was made in order to ensure a large region where the component of pressure normal to the track exceeded 5 psi; optimizing for a 5 psi overpressure would have reduced the effectiveness of the barraging weapons.

In order to maximize the normal component of the shock wave, it is clear that the weapons must be targeted to the sides of the track. This can complicate an attack in mountainous regions. The program *Eureka, the Solver*, published by Borland was used to fit the curves and to optimize the perpendicular offset from the track. Using the 5 psi normal criterion, a single 400 kt weapon can destroy trains over a total length of track of about 4.5 km when the weapons are offset a distance of about 1.7 km. Reducing the kill criterion to 3 psi normal and increasing the weapon yield to 500 psi raises the length of track covered by a single warhead to 6.4 km. The 3 psi criterion is clearly unrealistically small; the yield of a Soviet warhead, if its value is in fact correctly estimated by the US government, is classified.

The principal competitor to the MX rail-garrison system for a US mobile ballistic missile is the hardened mobile launcher (HML) containing a single-warhead Midgetman small ICBM . Earth moving equipment, the closest analog given in Glasstone and Dolan to armored vehicles, is approximately as rugged as a locomotive side-on. It seems reasonable to believe that armored vehicles could be built to accept 30 psi (the usual Air Force figure) and continue functioning. Present generation missile silos are in the 2,000-3,000 psi strength class; superhard silos are reliably believed to be capable of withstanding overpressures up to roughly 100,000 psi and (by some) also believed to be capable of remaining in place and vertical inside a blast crater (11)(see Chapter 11). These survival capabilities, then, give some criteria by which to distinguish between the three principal generic competitors for the next-generation land-based strategic missile: fixed-base in superhard silos; road/off-road mobile; and rail mobile.

Table 3. Damage criteria for basing modes

Superhard silos	100,000 psi
on or off-road mobile	30 psi
rail mobile	5 psi*

*Component of pressure perpendicular to the rails, but 2000 psi strong when in shelter at garrison (12).

The SS-18 missile has been tested with 10 real MIRVs being released and an additional 2 simulated releases. The SALT II Treaty limits the SS-18 (and all future ICBMs of either side) to no more than 10 re-entry vehicles on each booster. It is, nevertheless, possible that 12 re-entry vehicles are carried on some or all of the SS-18 force. The tests which have been carried out by the

Soviets might well have been sufficient to give them confidence in the reliability of missiles carrying 12 warheads. If the SS-18 carries its 12 warheads in a "two story" bus, six warheads per level, it is just possible for the maximum loading to be 14 weapons (13). The possibility that the SS-18 carries four warheads beyond the SALT limit is troubling, and complicates calculations of the effectiveness of the force. Nonetheless, such an additional burden must be compensated for by a reduction in the range of the missile below that at which it has been tested.

Despite the theoretical possibility that the SS-18 could carry more than 10 weapons, this paper will make the following explicit assumptions about the Soviet and US ICBM forces:

•The forces are constrained in number to the publicly available figures which would pertain under a START (strategic arms reduction) treaty (14).

•The Soviet Union would have no more than 154 SS-18s -- half of the present inventory -- and that each would be inspected to ensure that no missile carried more than 10 warheads (15).

At present the United States expects to have 500 Minuteman III missiles, each with three warheads, plus 50 MX missiles, each with 10 warheads, deployed in former Minuteman silos. The US has an additional 450 single-warhead Minuteman II missiles. The warhead total is, thus, 2450. If 50 additional Minuteman III missiles are retired and replaced by MXs, the US inventory would grow to 2,800 ICBM warheads. Perhaps more importantly, with this force the United States would retain 1,000 ICBMs, and hence 1,000 time-urgent targets which must be held at risk in a Soviet first strike. Although this number of ICBMs and launchers would be permitted under the US START proposal, it would allow only 2100 SLBM warheads. This is clearly too few to provide assured survival of an adequate number of sea-based missiles (an Ohio class SSBN carries 24 Trident II missiles, each with not fewer than 8 warheads, or 192 RVs per ship) on a large enough number of submarines. Congress has already authorized funds for 16 Ohio-class ships.

On May 3, 1988 the House Armed Services Committee adopted a sense of the Congress resolution introduced by Steven Solarz. This resolution suggested that a future START Treaty should not include any provision which would "result in a reduction of the total number of United States ballistic missile-carrying submarines below 20" (20). If all of those submarines are of the Ohio class or are similar to the Ohio class, and if all of the missiles are 8-MIRVed Trident IIs, the Navy would have 3,840 SLBM warheads. Only 1,060 weapons could be based on ICBMs under current START proposals.

Alternative strategies might include constructing the next batch of Ohio-class ships with only 18 or 20 missile tubes or offloading one or 2 warheads from each Trident II missile. Offloading would complicate inspection and verification of the number of warheads at sea, but would increase the range of the missile, and consequently the patrol area of the submarines. Reducing the Trident's war load to six RVs would reduce the number of missiles on a 20-ship force to 2,880 -- permitting 2,020 warheads on land-based missiles. It will cost roughly as much to operate 20 submarines carrying 24 6-warhead missiles as to operate the same number of ships, each armed with fewer, but more highly fractionated, missiles.

These force calculations are necessary in order to estimate the vulnerability of any one component of the land-based missile leg of the Triad. Even though 10-MIRVed MX missiles would be the most lucrative single class of land-based targets, the Soviet Union could not devote its entire arsenal to a strike on the MX component. These calculations indicate that it would take at least 1500 Soviet warheads to barrage the 500 rail MX warheads, a ratio of 3:1. It is doubtful that the Soviets could allocate even this large a fraction of their inventory to the rail MX component alone.

One knowledgeable observer (17) believes that the US will opt to retain 1,400-1,600 land-based ICBM warheads. If the existing 50 MX missiles are placed on trains, that would leave between 900 and 1,100 weapons for deployment on other ICBMs. The START rules encourage either single RV missiles or mobile single or multiple warhead missiles. Silo-based MIRVed ICBMs would become even more lucrative targets than they are today.

Figure 4. Artists concept of train in garrison. [US Air Force drawing.]

An attack on any one component of the US land-based strategic forces makes little sense unless the other components are hit simultaneously. Thus, a force of 900 silo-based missiles will divert 1,800 Soviet warheads from any attempt to barrage a mobile system. Under START rules, the Soviets would have no more than 3,100 warheads available for use against the rail-mobile system (18). Actually, a prudent planner would probably hold back a strategic reserve of roughly one third of his force, decreasing the number of ballistic missile warheads available for the first strike from 4,900 to 3,300. If 1,800 of those were used against silos, only 1,500 would remain for initial use against a mobile system. Not all of these will be loaded on highly accurate land-based ICBMs; indeed, the most accurate offensive missiles must be used against hardened silo targets.

We calculated above that a single SS-18 warhead can destroy trains over a 4.5 km (reckoning conservatively from the point of the offense) or 6.4 km (being defense-conservative) length of track. The probable Soviet strike force against a rail-mobile system is, therefore, capable of barraging between 6,750 and 9,600 km of track. Track is generated by a fleet of 25 trains dispersed from 17 garrisons at a rate of 6,625 km/hr according to Air Force calculations simulating real movement of the trains on the track. This rate of uncertainty is extremely high (265 km/train/hour) compared to the nominal 50 km/hr speed assumed for such trains. However, the US rail grid has (even in the middle of the Great Plains) a very large number of branch and intersection points. At each such point the train can go in either direction. When the train leaves garrison and its dedicated rail spur, it immediately picks up a factor of two in position uncertainty because it can be directed to the right or left. Thus, track generation at a rate of 265 km/train/hr is not an improbably high figure.

A length of track equivalent to that which can reasonably be barraged by the Soviet Union in a first strike (assuming START constraints) is thus generated in about 90 minutes. More conservative estimates still place the time to safety at 3 hours under START constraints.

Art Hobson demonstrates elegantly in Chapter 3 one basis for the missile designer's rule of thumb that in a barrage attack the *area* which can be cleared by a single missile is independent of the number of warheads on the missile. We now consider the effect of fractionation on the ability of an ICBM to attack a *linear* target.

Assume the throw-weight of the missile is fractionated among f warheads and that the quantity $f\,Y^{2/3}$ is a constant equal to C. The length of track, L, barraged by a single missile carrying f warheads of each of yield Y would be approximately equal to $k\,f\,Y^{1/3}$, where k is an appropriate dimensional constant:

$$L = k\,f\,Y^{1/3}$$

$$L^2/k^2 f = f\,Y^{2/3} = C$$

so

$$L^2 = C k^2 f$$

and

$$L \propto f^{1/2} .$$

Increased fractionation does, therefore, slowly increase the barrage capability of an ICBM force but at some cost in operational and testing difficulties. This avenue will be foreclosed in a START agreement by limitations on missile loadings, limitations which will be verified by some kind of inspection, as well as flight test restrictions as was done under SALT II. The START Treaty will automatically discourage the deployment of large numbers of highly fractionated missiles because it will limit warheads, not launchers.

Time to safety

The rail-mobile missiles will have to be dispersed (or "flushed") from the shelters in their garrisons upon receipt of strategic warning. If the US does not choose to announce that the trains have been flushed, there is no reason that the Soviets should know about the action for periods of hours (19). Suppose one assumes that the Soviets use two imaging satellites 12 hours apart, each reporting through a Soviet version of the US Tracking and Data Relay Satellite system. Each satellite makes fewer than 3 passes over the US each day, half of which are in darkness, for a total of 3 passes in daylight. The time interval between the passes is thus about 4 hours, and the mean time to detect the sortie of a single train, assuming the satellite is targeted on a garrison and sees evidence of the train in motion is about 2 hours.

The time to relay the information to the Soviet Union is vanishingly small, but the processing and analysis time at the far end is probably not less than 15 minutes, and more probably an hour (this includes constructing the image from the downlinked signal, looking at it, and making some kind of decision.). The minimum time one can expect to have before detection of sortie is thus about 2 or 3 hours, quite comparable to the time for the trains to reach a safe level of dispersal -- particularly when the 30 minute ICBM flight time is included. If the Soviets choose to execute a first strike against the rail-mobile system with short flight time SLBMs, the time delay *increases* because of the difficulty of communicating with the submarines.

The situation is really more favorable to the rail-mobile system than that, since a satellite cannot see terribly far off-nadir, 27° for the French SPOT remote sensing satellite, and hence cannot observe more than one quarter of the width of the US in any one pass. The Department of Defense estimates that the intelligence cycle time for the United States in a similar situation would be approximately 10 hours (20). The intelligence collection time can be reduced by adding satellites, but only at relatively great cost.

Raising the orbital altitude of the satellites increases the area which can be surveyed at one time much more rapidly than it increases the period of the satellite (21). Increasing the area in a single frame is not without complications, because, the resolution must decrease while the amount of data which must be transmitted each second increases. To gain back the resolution with more advanced sensor elements requires a great increase in the data-transmission capability.

Nevertheless, one former high official in the US intelligence community privately suggests that the minimum practical re-visit interval for *selected* points is about 90 minutes. That does not mean that the revisit time for *all* points can be shrunk to that low value; in particular, one cannot hope to scan enormous areas at that rate while retaining the 5 meter resolution needed to locate trains on pre-surveyed rail lines nor the (roughly) 1 meter resolution needed to identify a missile train moving along with other rail traffic.

Figure 5. Small model of a rail garrison MX missile launch car. [US Air Force photo.]

In addition, it is not obvious that a sortie will be detected unless the satellite is looking at the garrison just as the train leaves its horizontal silo, since an empty silo will closely resemble a full one. Furthermore, it is always possible to select the exact time at which the train leaves the silo and merges with the normal rail traffic for a moment when no observation satellite is over the horizon, since the ephemerides of most satellites are well known.

Because it is improbable that an attacker would know from national technical means the exact moment at which the trains were flushed onto the rail net, the attacker would always have to assume that the sortie took place right after his last inspection of the garrison. In effect, therefore, the attacker would have to barrage a length of track roughly corresponding to 90 minutes generating time even if the attack happened to land at the moment the trains left their shelters. The draft environmental impact statement for the rail garrison program indicates that the trains will be housed in massive shelters, perhaps having a 2,000 psi strength (22).

These same difficulties in satellite reconnaissance make finding or tracking trains on the rails extremely difficult, and hence maintain the invulnerability of a dispersed mobile system. The same assertion is true, of course, for on/off-road mobile systems, but the planned deployment schemes restrict their mobility to relatively small areas. In contrast, the track generated by rail mobile systems increases until the entire US rail net is filled --a total of nearly 200,000 route kilometers -- plus sidings, and yards. Even if the trains are required to avoid populated areas, the system will remain survivable.

Communicating with the trains in the pre-strike environment is relatively straight-forward since the US railroads have invested heavily in track-side communications systems, primarily fiber optics based. Additionally, the rails themselves form a transmission line. Richard Garwin, among others, has pointed out that after a strike the rails and fiber optics lines might well be severed. While it is not easy to sever rails with a nuclear weapon off-set from the tracks, the possibility cannot be excluded. In that event communication with the trains would be about as difficult as talking to HMLs which have been widely dispersed after the same kind of barrage. Because HMLs are stronger than trains, the barrage must be denser, so trailed fiber optics cables may not survive.

Calculations of the capability of HML-based missiles to survive an attack given either strategic or tactical warning will be found elsewhere in this volume. However, I would like to draw attention to some operational difficulties which affect the survivability of such missiles, particularly if they are stationed at existing silos which will continue to house ICBMs.

I had the opportunity to visit the Grand Forks Air Force Base missile deployment area recently. Most silos are located near to two-lane paved roads of reasonable quality. The roads are aligned on a nearly rectangular grid at intervals of 6 to 12 miles, with much poorer roads in between. Although one tends to think of the great plains as being flat, more like "planes," that is not the case. The Grand Forks missile deployment area is criss-crossed with stream beds, hills, rock outcroppings and the like. It includes large marshy areas as well. All of these make off-road mobile missiles less attractive operationally than they might seem from simply looking at a small-scale map.

Even more complicating for the notion of off-road mobile in this area, however are the long rows of trees planted during the dust bowl era to provide windbreaks. These have now grown to form almost impenetrable boundaries between fields. The actual area available for off-road mobile operations is almost certainly less than half of the apparent area. Worse, still, is the fact that the landscape, including the windbreak forests, will tend to channel the movement of HMLs along fairly predictable lines. Indeed, the survivability of HML mobile missiles in the Grand Forks area will more closely resemble that of rail-garrison missiles which generate track (i.e. linear uncertainty because they will be confined to the roads) rather than the hypothetical area generation used in the usual calculations of HML mobile survivability. Thus, the actual ability of a HML system to survive a short warning time attack (e.g. from submarine-launched missiles) might very well be far less than simple calculations would give. If SLBMs are also hard-target capable, basing HMLs at Minuteman (or other) missile silos might well give the attacker a welcome bonus in the form of mobile missiles destroyed as collateral damage to an attack on the silo.

Evaluation of the survivability of a real HML on/off-road mobile system, therefore, requires far more delicate calculations than we can make in this book. The survivability is a sensitive function of the road grid, the terrain, and the use of the land surrounding the precise points at which the missiles are based. The survivability of a rail-mobile system depends upon similar details, but the dependence is weaker because the rail net in the vicinity of the probable deployment sites is more nearly of a uniform quality than are the roads and off-road regions near existing Minuteman sites -- or even the desert deployment areas under consideration. Traffic control and congestion is also a problem, but the MX office indicated that the railroads were willing to allow the Air Force to take over control of rail traffic during an emergency.

Nevertheless, it is apparent that both the rail-garrison and HML systems are survivable if the National Command Authority gives the order to disperse based on strategic warning (the kind of warning which the "MAGIC" intercepts of Japanese communications should have given the United States on December 6, 1941) rather than waiting for the tactical information that a Soviet attack is under way. Even assuming only tactical warning the rail-garrison system can be made more survivable than conventional silos, particularly in a START-constrained environment. This is an important conclusion derived from the assumptions that the garrison is, itself, approximately as strong as a conventional silo and that the Soviet intelligence cycle time plus the missile flight time is at least 30-90 minutes so that a significant length of track must always be barraged (23).

The choice of a modernized land-based missile system must depend, therefore, on factors other than simple calculations of survivability. These can include environmental impact, crisis stability, arms control verification considerations, and most importantly, cost.

Crisis stability of mobile systems

All analysts agree that strategic crisis stability is important. The Scowcroft Commission put it succinctly: "Stability should be the primary objective of both the modernization of our strategic forces and of our arms control proposals." The definition of stability can be made quantitative, but the essential intuitive definition is that the situation is stabile against a first strike when there is little or no difference in the cost to each side between striking first and riding out the other side's attack before retaliating. If the cost incurred by going first (i.e. the damage done by the other side's retaliatory strike) is significantly (24) less than the cost of going second, that side may feel pressure to strike first in order to avoid the worse consequences which would follow from an enemy's first strike (25).

Only when each side perceives that there is no disadvantage to absorbing a first strike is there any chance of preventing central nuclear war. It is, therefore, essential to provide survivable strategic weapons which can hold at risk a large fraction of the other side's most valuable targets. Indeed, the situation is most stable if the fraction of the enemy's value targets which can be destroyed is very nearly the same in a retaliatory strike as it would have been from a first strike. This would be the case for fully survivable strategic forces.

The process of generating forces and bringing them to the alert can, therefore, communicate two messages. It obviously puts our adversary on notice that we are serious -and requests a deescalatory response. When mobile forces are dispersed, however, the message is somewhat more complicated. The principal signal from the dispersal of mobile missiles is that they are being made more survivable and less targetable. In dispersed deployment they do not present any more of a threat to the opponent's forces than they do in garrison or, in the case of HMLs containing small ICBMs, in peacetime deployments. Therefore, acting to sortie mobile missiles is not threatening. To the extent that the performance of missile guidance systems is temporarily degraded by movement, placing mobile missiles on alert may actually reduce the threat they pose to enemy targets. In a situation of approximate crisis stability, therefore, flushing rail-garrison trains or HMLs from their peacetime deployment sites reduces the cost to the United States of going second while reducing the benefits to the Soviets of going first. It is, therefore, inherently stabilizing.

The question of strategic warning has been addressed by many commentators (26). Such warning should be available. According to SAC officers who briefed associates of the Carnegie Endowment for International Peace, the peacetime deployments of Soviet strategic forces are distinctive: 90 per cent of the SSBN force remains in port; there are no armed alert bombers; and tanker aircraft are not positioned to support strategic bombers.

Deviations from these deployments should be clear, but the critical question is whether or not the United States would take definitive action based on warning. If that action were perceived in both countries as unambiguously stabilizing, the chances of a timely response might be assessed as fairly high. On the other hand, General Douglas MacArthur had adequate strategic warning of the Japanese attack on the Philippines (because it occurred many hours after Pearl Harbor was bombed). Nevertheless, most of his aircraft were destroyed on the ground lined up in peacetime formations.

If the international situation is deteriorating rapidly, one side may be tempted to strike first in order to eliminate the other's mobile missiles before they can reach safety. If elimination of any one component of the American strategic force mix could reduce the cost to the Soviet Union of going first so much that a first strike appeared more rational than waiting and hoping to avoid a nuclear conflict, that would be an indictment not of mobile missiles, but of the total US strategic force doctrine.

In order to avoid precisely this situation, the United States and the Soviet Union each deploy several different kinds of strategic nuclear forces, each having a different failure mode. Mobile land-based forces form only one part of a complex of strategic forces. One advantage survivable mobile forces have over missiles in silos is that placing mobile missiles on alert is stabilizing, while placing vulnerable forces on alert indicates preparation for a first strike.

Arms control implications

Some argue that rail-mobile missiles complicate the verification of arms control agreements. These points are raised: rail-mobile missiles can roam all over the country, be hidden in tunnels and buildings, or be concealed in civilian-appearing trains. But rail-mobile missiles are just that -- confined to the railroad net. Rail lines are easily recognized features when seen in satellite photography, even at the 10 meter resolution provided by the French SPOT 1 satellite. The Soviet rail net is particularly easy to inspect, since it is a skeleton whose back bone is formed by the Trans-Siberian and BAM rail lines stretching across 7 time zones, but which has few ribs running perpendicular to the spine. Trains on tracks are identifiable in high-resolution imagery, but such pictures necessarily cover only fairly narrow angles of view. Hence, they cannot be used effectively to search for trains which have been ordered to sortie from garrison.

Trains cannot be hidden for long periods in tunnels; concealing a missile there would block an entire rail line, and construction of special tunnels would probably be observed. Trains cannot circulate at random, but must follow schedules -- particularly on a single-track main line such as the Trans-Siberian -- since trains in opposite directions can only meet where sidings exist. Rails rarely enter buildings; they stop at loading docks. The relatively few points at which tracks enter buildings in a suspicious way could be inspected if an arms control agreement limited the number of mobile missiles and their deployment sites.

Finally, the facilities at which missiles are mated to rail cars will be very distinctive, easily recognized in satellite photography. Relatively straight forward analysis based on imagery of such a facility can provide good estimates of its maximum through-put, and hence an upper limit to the number of deployed rail-mobile missiles. The number of such facilities in each country can readily be made a part of any future arms limitation or reduction agreements.

Rail-mobile missiles confined to garrison in peacetime present less of a verification problem than do missiles loaded in much smaller vehicles -- HMLs -- and autonomously mobile at all times. Indeed, the heavier the mobile missile, the more detectable its infrastructure will be, and hence the more confidence one can have in a verification regime. Rail garrisons also present fewer locations which might have to be monitored on-site by Soviet (respectively American) inspectors, and hence reduce the load on the inspection regime. If, however, a moderate number of unpermitted and highly-MIRVed rail-mobile missiles are hidden from the inspection regime, they would present a greater threat than the same number of single-warhead Midgetmen (respectively SS-25s).

A typical MX train will consist of one or two locomotives, one launch control car, two missile cars, and two personnel cars. It is, therefore, considerably smaller than a typical American train. Those few trains which, in peacetime, need to be out of garrison (for example to take missiles for routine service) will, therefore, be extremely distinctive. In time of war, of course, additional cars can be added to reduce the signature of the missile train and make its detection by observation satellites much less probable.

It is desirable to have two locomotives for redundancy and to strengthen the train at both ends. The addition of a tank car carrying diesel fuel will permit the locomotives to travel for many days. Since crew sleeping and messing facilities would require only a modest adaptation from existing Pullman and dining cars, and since provisions for several weeks could be easily accommodated, the endurance of such a train might be comparable to that of a nuclear submarine on patrol.

It is worth pointing out that unclassified SPOT satellite reconnaissance of the SS-25 site at Yurya shows that the missile deployment is confined to a relatively small operating area consisting of a central road with spurs to prepared launch sites. The road does not appear to connect in any easy way with the civilian road net in the region (27). This mode of operation is significantly different than any proposed for the US Midgetman.

In a START-constrained environment, a force of single-warhead silo-based missiles is presumptively safe from attack because the number of targets can equal or exceed the number of warheads which can be aimed against the force. For the United States this would require the construction of a large number of additional hardened silos if we desired to approach the 4,900 limit on total ballistic missile warheads and decided not to retain MIRVed missiles. The missile of choice for the silos might well be the Midgetman, now under development or the hypothetical Minuteman IV. An alternative solution would be to include a moderate number of two or three warhead missiles, reducing the number of targets presented to the Soviets but reducing the cost of the force.

A third solution would be to phase out the existing 450 Minuteman II single warhead missiles and to replace them with Midgetman missiles as the existing force ages in the 1990s. Then the use of rail-garrison MX missiles to enlarge the target base the Soviets would be required to barrage in a first strike would make eminent sense. The rail-garrison missile ought to be preferred over the HML system on the grounds of: reduced environmental impact; slightly diminished public interface problems; comparative ease of arms control verification; and cost.

Public interface and environmental concerns

The HML system would have to be deployed on enormous tracts of federal land in the Southwest or off of federal land and in enlarged and modified Minuteman silo enclosures. Deployment at Minuteman sites is apparently less expensive, and is currently favored by the Air Force. Each enclosure will house two missiles and their crews, who would have to be continuously present to gain the "bolt out of the blue" survivability advantage alleged for the Midgetman system. The addition of the HML enclosures and crew quarters will significantly impact the use of what is now generally farmland.

Even routine service of the HMLs would probably require them to be driven on public roads, with their missiles, to reach maintenance areas. This is particularly true if they are deployed at existing Minuteman III sites which are only served by public roads.

The rail-garrison system, in contrast, will remain wholly on SAC controlled government land and will be housed in shelters on bases which already are home to nuclear missile forces. The trains and their missiles can be maintained, serviced, and exercised without ever leaving their bases and a short stretch of dedicated rail track constructed therein. The total amount of conventional construction needed to house the rail-garrison system will be far smaller than that needed to house the HML force. The draft environmental impact statement (28) gives quantitative estimates of the size of the rail mobile system and its infrastructure.

Some critics of the rail-garrison system suppose that public reaction in a crisis might prevent the president from dispersing the missile trains onto the rail network. While this possibility cannot be entirely discounted, it can be set in context. First, the possibility and significance of dispersal will be discussed and debated extensively in Congress before the system is constructed; this should serve as a time for public education on two salient points -- that dispersal is a way of increasing crisis stability, and that the missiles are safe unless armed for launch. Second, in a situation in which the Soviets are perceived to be threatening the United States directly, the American public is apt to be extremely supportive of measures taken by the president to avert war. It is unlikely that there will be concerted public action to hinder the dispersal of the trains (or of small ICBMs onto open land and public roads).

Cost

It is generally conceded by partisans of both rail-garrison and HML deployment that the cost of the rail system will be on the order of $10 billion while that of a mobile Midgetman system will be roughly $40 billion over a 10-year life cycle for similar numbers of warheads. The differences in the price tags originate in, primarily, two areas. The HML system uses many more missiles for the same number of warheads since it is not MIRVed, and the HML system uses far more expensive launch vehicle since the HMLs must be hardened to a considerable extent, must carry crew and communications equipment, and must be mobile. The differences in crew salaries alone is a significant factor.

It may also be unrealistic to consider a 10 year life-cycle. Most American strategic weapons systems have been retained in service for far longer -- up to 30 years are now scheduled for the Minuteman II and III. While the amortized differences in capital costs would appear to be less of a factor over a longer life-cycle, the differences in manpower will loom far more important than they would for a 10 year period.

Rail equipment, in contrast to HMLs, is cheap. Acquisition of equipment and construction of facilities for a 25 train, 50 missile, MX rail-garrison force has been estimated at $10.4-$12.1 billion in 1988 dollars (29). Used but adequate locomotives can be bought on the open market for $100,000; new locomotives cost around $2 million. The incremental cost per train, not including the missiles, and not amortizing research and development funds over trains beyond the planned initial deployment of 25, is roughly $160 million.

The rail system needs no more than 4,300 people to operate it, year round (30); the HML system requires not fewer than 5,000 crew members in the HML cabs as well as many more for support, command, and control (31). Excluding the salaries for peripheral personnel, the rail-garrison system will cost $200 million/year for operations and support.

Summary

The rail-mobile system possesses four distinct advantages over all of its competitors:

•Pre-attack communications with the rail system can take place either through the rails or through the fiber-optics system which parallels most main-line tracks. In this case "pre-attack" means the period until warheads land.

•The location of a car on the rail lines can be routinely ascertained with existing rail systems (not satellites) to better than one meter, thus permitting accurate recalibration of the missile guidance systems after a movement (32).

•The endurance of a dispersed rail system can be very high. A train using two locomotives and a diesel fuel car can stay on alert for weeks at a time if necessary.

•The cost of a rail mobile system, calculated either in absolute terms or per warhead, is far less than that of an on/off-road mobile system,while it is far less vulnerable than missiles deployed in fixed-silos.

In an era of shrinking real defense budgets it is difficult to urge the choice of the more expensive of two otherwise competitive strategic systems. Because a START Treaty will probably place limits on warheads rather than on nuclear delivery vehicles, it discourage highly MIRVed systems unless those systems are effectively untargetable -- which the rail-garrison MX system is under most circumstances. The federal budget will force reconsideration of the admitted strategic advantages of manpower-intensive and "booster-intensive" (33) systems, which describes the Midgetman in HMLs.

The single-warhead Minuteman II missiles and three-warhead Minuteman III missiles now deployed in silos are an useful component of a START-constrained strategic missile force. They stress the capabilities of the total Soviet missile force, ICBM and SLBM, so that the US land-based force cannot be attacked with confidence. For this reason, it remains reasonable to pursue the development of a new single-warhead ICBM *for deployment in silos* (and possibly on trains) at relatively low -- and certainly predictable -- cost. Such a missile need not have hard-target kill capability, and might well be equipped with a warhead significantly smaller than that on the Minuteman II -- our only remaining single-warhead strategic ballistic missile. Such a missile could be called either Minuteman V or Midgetman.

References and notes

1. Submarine damage criteria, and methods of destroying submarines other than by explosives in direct contact with their hulls, are both complex and highly classified. In general, however, it can be said that damage is inflicted not by crushing the hull but by accelerating the ship and dislodging machinery.

2. Concealment can be obtained by combining the missile train with a number of dummy freight cars so that it mimics the signature of ordinary rail traffic. Such camouflage will not be perfect because at least two cars in the missile train will remain fairly distinctive.

3. Samuel Glasstone and Phillip J. Dolan, *The Effects of Nuclear Weapons*, Third Edition. US Departments of Defense and Energy, Washington, D.C. (1987), p. 193. Hereafter cited as *Effects*.

4. *Effects*, p. 191.

5. *Effects*, p.. 192.

6. Lt. Col. Tom Maxwell, USAF, head of rail-garrison MX project, private communication.

7. *Effects*, Figures 5.95 a and b, p. 195. Whether these figures are fully applicable is, however, questionable. The tested planes were World War II vintage B-17s and were tethered to the ground so that they could not move with the blast wave. It would not be unreasonable to speculate that modern jet bombers, when airborne, in particular, are significantly harder.

8. Physically, however, some insight is gained by recognizing that, dimensionally, energy density and pressure are equal, in appropriate units. An explosion of given yield (energy release) Y (in kilotons) fills and expanding sphere of radius r with energy. The energy density (or equivalently, pressure), at a given radius should thus vary roughly as Y/r^3. Alternatively, the radius at which a given overpressure occurs for weapons of varying yields scales as $R=R_1Y^{1/3}$.

9. See, for example, *Effects*, chapter 3, and Defense Nuclear Agency documents such as *Weapon Effects CROM Reference Handbook*, DNA-EH-84-01-V2, not generally available to the public, but nevertheless unclassified. Distribution is restricted because the data are considered "critical technology;" export of the information is restricted by the Arms Export Control Act or Executive Order 12470.

10. The precise yield of the warheads in SS-18 Mod 4 reentry vehicles is, of course, unknown in the US. Values ranging from 350 kilotons to 750 kilotons have, at one time or another, been quoted. The 1987-88 edition of the IISS's *Military Balance* gives, for example, a yield of 500 kilotons. Since the distance at which a given overpressure is produced scales only as the cube root of yield, the difference between assuming 500 kilotons and assuming 400 kiloton yields is only about an 8% difference in radius. Because the barrage effect against railroad track depends on radius, not area, of damage, the difference is indeed small. The survivability of a rail-mobile system may depend more sensitively upon assumptions about the number of warheads on the barraging missile (see below).

11. This does the superhard silo little good in a two-on-one attack since a second weapon detonating anywhere in the vicinity will knock over the exposed silo.

12. The strength of the garrison structure has been given as 2000 psi by Air Force officers interviewed by the author. The draft environmental impact statement is silent on the question. Early public statements made by Air Force officials indicated that the garrisons might only survive relatively small blasts, i.e. on the order of 100 psi.

13. The SS-18 was tested with ten real warheads and two simulated warhead releases. Some in the US intelligence community believe that the conduct of the tests demonstrated that the missiles could, in fact, be loaded with 12 RVs, stacked in two layers of six each. Consider a ring of six warheads; even if all six warhead touch one another, there is precisely enough space in the middle of the layer to accept a seventh RV. If the six warheads have a significant amount of space between them, it becomes even easier to fit a seventh into the center. For this reason some analysts fear that the SS-18 may carry as many as 14 warheads.

14. The US has called for a limit of 4900 ballistic missile warheads, with a sublimit of 3000-3300 of those permitted on ICBMs. The Soviet Union has called for equal warhead sublimits on ICBMs and SLBMs (see *Arms Control Update*, May 1988, p. 7, US Arms Control and Disarmament Agency). The communique from the Moscow summit did not indicate any change in the position of the two sides as regards the question of ICBM and SLBM sublimits.

15. The number of SS-18s to remain after a START treaty has been agreed upon, as has the number of weapons which may be deployed on those missiles.

16. Michele A. Flournoy, "START thinking about a new US force structure," Arms Control Today, Jul/Aug 1988, pp. 8-14.

17. Michele A. Flournoy, private communication.

18. On the plausible assumption that each silo-based missile must be attacked with two warheads to ensure a very high probability of destroying the target. The two warheads must, further, be launched from different missiles, a procedure known as "cross-targeting," so that if one missile fails its assigned targets are otherwise covered.

19. It might be possible for the Soviets to plant spies near the gates of the rail garrison or to locate sensors on or near the tracks. While the use of *Spetznaz* troops cannot be excluded, it would present difficult operational problems for the Soviets, and would be extremely embarrassing if any such teams were detected prior to war breaking out. The question of sensors is more difficult to assess since they could, in principle, be quite small, relying upon low-power radio transmission to locations where clandestine satellite up-links could be placed. Nonetheless, these sensors would have to be so well concealed that they were not detected in the course of regular

inspections of the rail lines for many kilometers in each direction from the point at which the dedicated rail lines (on the missile base) join the main line.

20. Private communication with knowledgeable observers who decline to be identified.

21. The area which can be seen for a given off-nadir pointing capability increases as the square of the average altitude above the surface of the earth; the period of a satellite in circular orbit varies as the 3/2 power of the radius of the orbit itself--including the 6400 km radius of the earth.

22. See *Draft Environmental Impact Statement: Peacekeeper Rail Garrison Program*, US Air Force, Washington, D.C. (1988).

23. The flight time of an ICBM is 30 minutes; that of an SLBM is 15-20 minutes on a conventional trajectory. In calculating the size of a barrage against a mobile target, the offense must always assume that the dispersal time is at least as great as the flight time of its missiles plus the time since the last satellite inspection of the target to be attacked.

24. As judged by the participant. "Significant" is not readily quantifiable; the benefit of having no nuclear war is hard to overestimate.

25. Much of this analysis was originally performed by Lt. Gen. Glenn Kent (USAF-Ret.) of the RAND Corporation, and is taken from the RAND unclassified briefing "First strike stability: a criterion for evaluating strategic forces" (unpublished).

26. e.g. R. James Woolsey in his address to the Stanford University Center for International Security and Arms Control, 14 Oct 1988.

27. Picture courtesy of Space Media Network, Stockholm and in the possession of the author. Copyright CNES and SMN.

28. op. cit.

29. Private communication, Lt. Col. Tom Maxwell, USAF, Rail-garrison system project officer.

30. Table S-1, "Direct employment---," *Draft Environmental Impact Statement, Peacekeeper Rail Garrison Program*, US Air Force, Washington, D.C. (1988).

31. The calculation goes as follows: 500 HMLs, each with a crew of 2, require 1000 people on duty at all times. To man a single post continuously in peacetime requires 5 people for each post when 8 hour shifts, 40 hour weeks and 30 days annual leave plus holidays are included. Therefore, not fewer that 5000 billets will be required just

to fill the crew cabs of the HML system. The actual number could easily turn out to be double that when maintenance, communications and command personnel are included.

32. According to interviews with railroad officials.

33. Because each booster will require its own guidance system, launching equipment, communications and crew facilities.

Part V Chapter 9

Engineering of missile silos

John R. Michener

Introduction

Missile silos are designed to protect their contents, ballistic missiles, from the effects of adjacent nuclear weapon explosions. They therefor are required to protect their contents from the shock, overpressure, electromagnetic radiation of all frequencies, and neutron pulse generated by the nuclear weapon upon detonation.

Protection from radiation and electromagnetic effects

Missile silos are typically massive structures of steel and concrete that are buried in the ground. It is relatively easy for such structures to provide substantial protection from blast induced electromagnetic radiation, neutrons, and air drag. Continuous metal skins on the inside and outside of the silo spaced by a substantial thickness of concrete provide a doublewall Faraday cage that is very effective for such protection.

The substantial thickness of concrete, steel, and earth provide good protection from the gamma rays and neutrons released by the bomb and its radioactive products (1). While dry ground does not have a strong interaction with the prompt neutron pulse emitted by the bomb, damp ground and concrete interact quite strongly due to their water content. 30 cm of common concrete reduces the neutron energy flux by about a factor of 10 and the gamma ray dose by a factor of 4. 10 cm of steel reduces the gamma ray exposure by a factor of 6 (a strong silo is likely to have on the order of 1 m of concrete and in excess of 10 cm of steel in its wall and cover). Loading the concrete with high atomic mass material such a barium sulfate as well as neutron absorbers such as boron containing materials substantial increases the interaction with gamma rays and neutrons, increasing the shielding factor for both these radiations. Since the missile silos are buried and flush with the ground, they lack a significant aerodynamic cross-section, and are not significantly exposed to damage from the dynamic air drag and entrained debris that can otherwise damage exposed surface structures. The primary damage mechanisms that have to be treated are shock damage to the missile and mechanical crushing of the silo structure.

Radiation hardened electronics are able to survive radiation exposures in excess of 10^5 Rad and may be able to survive exposures to 10^6 Rad, depending upon the circuit. While radiation releases by attacking weapons may exceed 10^8 Rad, the massive walls of the silo and earth provide substantial shielding. A shielding layer of 1 m of concrete and 10 cm of steel would reduce the neutron flux by >600 X and the gamma flux by >2000 X, without the use of additional absorbers in the concrete. Additional neutron shielding can easily be provided by the use of a layer of borated polyethylene within the silo cap and/or a m or so of loose earth covering the cap. The availability of radiation shielding in silos means that the survival of the electronics critical to the missiles mission is not at risk and radiation damage is not a limiting factor in missile survivability.

Shielding from overpressure

The air blast from a nearby nuclear explosion is characterized by an almost instantaneous rise time and a decay time constant (10%) of 2 to 5 msec. The short time of the applied pressure makes the proper treatment of this problem one of dynamics rather than one of statics. The mechanical properties of materials upon such time scales are considerably superior to their static properties. When loaded for such short periods dislocations do not have much time to move or multiply and cracks are harder to nucleate and grow. The dynamic treatment of a silo structure is a difficult problem and much of the necessary information upon material properties at these high strain rates is not readily available. The use of static analysis and associated steady state materials properties yields a very conservative analysis. In the analysis of silos that follows static assumptions and materials properties will be used. The strengths of structures that this analysis yields are highly conservative. Discussions with individuals familiar with the dynamic calculations suggest that the static analysis understates the structure survivability by several times.

Crushing of the missile silo and its cover is the mechanism that most limits missile protection from nearby nuclear weapons detonations. This class of destruction mechanisms may involve buckling of protective walls, material failure of the silo and protective structures because of the applied pressure, or failures due to nonuniform blast loadings causing high bending loads that induce local failures and the resultant destruction of the protective structure. Thin walled structures (2) are particularly vulnerable to buckling failures since their resistance to local bending is relatively low. While relatively thin walled structures are necessary for mobile protective structures, thick walled structures being too massive to move, they are not used for in-ground silos, which can be far more massive.

Consider the crushing of an idealized missile silo composed of a thick walled cylinder with spherical end caps which share the inner and outer radii of the main cylindrical section. Initially the material of the silo will be assumed to be homogeneous and isotropic, simplifying the analysis. Consider first the response of two idealized geometries: a thick walled cylinder of inner radius a and outer radius b, and a thick walled spherical shell with the same inner and outer radii subject to internal and external pressures. The solutions of these problems are well known in the field of elastic mechanics (3). A relatively simple derivation of the stress solution for a thick walled cylinder is presented in Appendix 1. This solution is:

$$\sigma_{\theta\theta} = P_i \frac{(\frac{b}{r})^2+1}{(\frac{b}{a})^2-1} - P_0 \frac{1+(\frac{a}{r})^2}{1-(\frac{a}{b})^2}, \qquad \sigma_{rr} = -P_i \frac{(\frac{b}{r})^2-1}{(\frac{b}{a})^2-1} - P_0 \frac{1-(\frac{a}{r})^2}{1-(\frac{a}{b})^2}$$

The corresponding solution for the strain is (case of plain stress):

$$u_r = \frac{(1-\upsilon)(a^2P_i - b^2P_0)r + \dfrac{a^2b^2(P_i-P_0)(1+\upsilon)}{r}}{E(b^2-a^2)}$$

where the external pressure is P_o, the internal pressure is P_i, r is the radial distance from the center of the cylinder, a is the inner radius of the cylinder, b is the outer radius of the cylinder, E is the elastic modulus of the material, and n is the Poisson ratio of the material.

In the limiting case that b >> a, the expressions for the stress simplify considerably:

$$\sigma_{\theta\theta} = P_i(\frac{a}{r})^2 - P_0[1+(\frac{a}{r})^2] \qquad \sigma_{rr} = -P_i(\frac{a}{r})^2 - P_0[1-(\frac{a}{r})^2]$$

The stress solution for the case of the spherical shell has a similar structure

$$\sigma_{\theta\theta}=\sigma_{\phi\phi}= P_i\frac{(\frac{b}{r})^3+2}{2[(\frac{b}{a})^3-1]} -P_0\frac{2+(\frac{a}{r})^3}{2[1-(\frac{a}{b})^3]}, \qquad \sigma_{rr}= -P_i\frac{(\frac{b}{r})^3-1}{(\frac{b}{a})^3-1} -P_0\frac{1-(\frac{a}{r})^3}{1-(\frac{a}{b})^3}$$

as does the corresponding solution for the strain

$$u_r=\frac{(1-2\upsilon)(a^3P_i-b^3P_0)r - \frac{a^3b^3(P_0-P_i)(1+\upsilon)}{2r^2}}{E(b^3-a^3)}$$

As in the case of the cylinder, the expressions for stress simplify considerably when b >> a

$$\sigma_{\theta\theta}=\sigma_{\phi\phi}= \frac{P_i}{2}(\frac{a}{r})^3 - \frac{P_0}{2}[2+(\frac{a}{r})^3] \qquad \sigma_{rr}= -P_i(\frac{a}{r})^3 - P_0[1-(\frac{a}{r})^3]$$

It can be seen that in this limiting case, the stress at the inner wall of the cylinder is twice that of the pressure applied to the silo (when no internal pressure is applied). Similarly, the stress at the inner wall of the spherical shell is 1.5 times the pressure applied to the shell. As the wall thickness is decreased, the ratio of the internal to the external stresses increases. Consider the case of a silo with a wall thickness equal to its internal radius, b=2a: At the inner wall of the cylinder, r=a, and so

$\sigma_{\theta\theta} = -2.67P_o,$ $\qquad \sigma_{rr} = 0$

while at the outer wall of the cylinder, r=b=2a, and

$\sigma_{\theta\theta} = -1.67P_o,$ $\qquad \sigma_{rr} = -P_o$

At the inner wall of the spherical shell, r=a, and so

$\sigma_{\theta\theta}= -1.71P_o,$ $\qquad \sigma_{rr} = 0$

while at the outer wall of the spherical shell r=b=2a, and

$\sigma_{\theta\theta} = -1.21P_o$ $\qquad \sigma_{rr} = -P_o$

Comparison of the behavior of the thick walled cylinder and the spherical shell reveals that the spherical shell is stiffer and stronger than the cylinder. Failure of a composite structure composed of hemispherical end caps on a cylinder can be expected to occur within the cylinder. In the case of a silo, the top cap must be removable to allow the firing of the enclosed missile. The requirement that the top be easily lifted off means that while the cap and the cylinder can transfer mechanical loads to one another, the greater stiffness of the cap can not be efficiently utilized to stiffen the top region of the cylinder.

Conventional high strength concrete has a compressive yield strength of from 200 to 300 atmospheres. Easily available techniques and materials are sufficient to increase the compressive yield strength to roughly 1000 atmospheres at little cost (4-9). The corresponding tensile strength is on the order of 8-10% of the compressive strength. In the geometries considered above the stress is not uniaxial: at the inner wall of the cylinder there will be stresses in the theta direction as well as the z direction (transferred from the load on the end caps). In the hemispherical caps there will be stresses in the theta and phi directions. In both cases the radial stress is zero at the inner wall (a boundary condition). In a brittle material, such as concrete, failure is typically taken to occur when the principle stresses exceed either the compressive or tensile stress limits of the material (10). The stresses calculated above are the principle stresses and failure can be assumed to occur when these values exceed the strength of the material in the inner wall.

While the primary loading of the silo structure will be in compression, the passage of the shock wave, nonuniformities in loading, and applied bending moments will result in the generation of tensile stresses in the silo material. An unreinforced concrete structure is very vulnerable to such stresses, since the tensile failure stress for concrete is a small fraction of the compressive failure stress. Reinforcing steel has a tensile strength in excess of 5000 atmospheres and a stiffness (elastic modulus) at least 6 times that of standard concrete (high strength concrete has a higher elastic modulus than does standard concrete). The appropriate placement of several volume percent of reinforcement rods in the silo structure can significantly increase the resistance of the structure to compression as well as allow the material in the structure to survive tensile loads comparable to those required to cause compressive failure. The presence of a stiffened inner silo wall establishes a compressive radial stress in the concrete surrounding the inner region. This results in an effective hydrostatic stress in the surrounding concrete, increasing its resistance to failure. As a result of these properties and design characteristics, the construction of silos using conventional concrete and steel construction to survive overpressures on the order of 100 to 200 atmospheres is straightforward.

Multishell structures

The calculations in the preceding section indicated that silos constructed of cylindrical and spherical sections are most likely to fail in the cylindrical region (11). The following calculations will therefore be for multiwalled cylindrical structures, since the same multiwall structure in the end caps would result in lower stresses than would occur in a corresponding cylindrical structure.

Consider a three wall cylindrical structure subject to the conservative assumption that the bond strength between the layers is negligible. The presence of substantial bonding between the layers, as would be expected in the course of good construction practice, would aid in the transfer of loads between layers and would minimize local stress concentrations. The basic structure of interest for this calculation is two concentric steel shells with the space between them filled with concrete (which could be reinforced as well). For reasons of economy and construction ease, the intermediate concrete thickness will be taken to be much thicker than the steel shells.

The presence of the massive concrete shell stiffens the steel shells against buckling, which would otherwise easily occur with such relatively thin walled shells. When the contact between layers is assumed to be frictionless the contact stresses are hydrostatic (no shear is transmitted). This allows the calculation of the behavior of the composite as a superposition of the behavior of thick walled shells subject to internal and external pressures. The inner wall of the inner shell has no internal pressure while the outer wall of the outer shell is subject to the applied pressure. The deformations of the shells at their respective boundaries are to set equal to one another, creating expressions coupling the contact pressures to the applied pressure through the geometry of the system and the elastic properties of the respective layers. Once the contact pressures are known the stresses in the materials are easily determined. Solution for the contact pressures is straightforward but exceedingly cumbersome if done manually, but is simple using symbolic manipulation programs such as MACSYMA. Performing these manipulations and evaluating the resultant expressions for materials and geometries of potential interest using MACSYMA, the values in Table 1 were generated.

Examination of these results indicates that the presence of modest thicknesses of steel/iron substantially reduces the stresses present in the concrete at the expense of localizing substantially increased stresses in the metal shells. The presence of very rigid inner shells induces considerable hydrostatic stresses in the concrete, considerably reducing the effective non-hydrostatic loads in the concrete. The distribution of stress between the different shells is determined by the relative elastic moduli of the shells. Shells with relatively high moduli are more rigid and are forced to carry a higher portion of the stress than they would carry if the moduli were more equal. If the moduli of the shells are equal the stress situation would be identical to that of a uniform thick shell.

The addition of a substantial volume fraction of steel wire or amorphous iron-boron ribbon to the concrete increases the average elastic modulus of the concrete and reduces the stresses in the cement matrix. It also significantly increases the fracture resistance of the concrete (12) and its radiation absorption cross-section. This process is discussed in Appendix 2.

The use of high strength concrete systems with wire reinforcement is desirable because of the stronger bonding and lower defect density obtained with the high strength concrete systems. If substantial steel wire loadings are used in the concrete, the failure load of the resulting composite can be substantially increased, but the concrete/steel composite also carries more of the applied load. A schematic of a hard silo structure from the open literature (12) is provided in Figure 1. In this structure massive wire reinforced high strength concrete is used in the silo walls. Table 2 presents results for the average stresses in steel-concrete composite-steel multiwalled structures as well as the stresses in the cement matrix of the composite.

Table 1. Relative induced stresses in a multilayer cylinder by an applied external pressure.*

location	inner shell thickness (cm) & mass (x 1000 kg)	outer shell thickness (cm) and mass (x 1000 kg)		
		2.5 (46)	5 (92)	10 (180)
σ,inner shell	2.5	12	11	9.9
σ,concrete core	(21)	2.0	1.9	1.7
σ,outer shell		7.2	6.7	6.0
σ,concr.-hydrostatic stress		1.7	1.6	1.4
σ,inner shell	5	10.3	9.7	8.6
σ,concrete core	(43)	1.7	1.6	1.5
σ,outer shell		6.6	6.2	5.6
σ,concr.-hydrostatic stress		1.3	1.2	1.1
σ,inner shell	10	8.1	7.7	7.0
σ,concrete core	(88)	1.4	1.3	1.2
σ,outer shell		5.5	5.4	5.0
σ,conc.-hydrostatic stress		.68	.64	.58
σ,inner shell	20	6.0	5.7	5.3
σ,concrete core	(186)	1.0	.99	.91
σ,outer shell		4.8	4.6	4.2
σ,concr.-hydrostatic stress		.12	.12	.11

* All stresses are calculated at the inner wall of the respective layer because this is the location of the peak stresses in the layer. Mass calculations are for low alloy steel. The length of the cylinder is assumed to be 15 m, the inner radius is 1 m and the outer radius is 2 m. Mass figures include the hemispherical end caps. The applied stress = 1.0. Standard concrete was assumed to be used, resulting in a 6:1 elastic modulus ratio between the steel and the concrete. Current dimensions for Midgetman are r<1.2 m, length<14 m.

High strength ductile cast iron has an ultimate compressive strength on the order of 6000 atmospheres. The compressive failure loads of high strength, high toughness, plate steels may approach twice this value. Ultrafine grain, ultrahigh carbon steel (carbon content of about 1.5%, a material that is not in current engineering use, can have compressive failure loads well in excess of 12,000 atm. While ordinary high strength concrete has a failure load on the order of 200 to 300 atmospheres, the compressive yield strength can be easily raised to around 1000 atmospheres by control of the composition and mixing ratios (5,6).

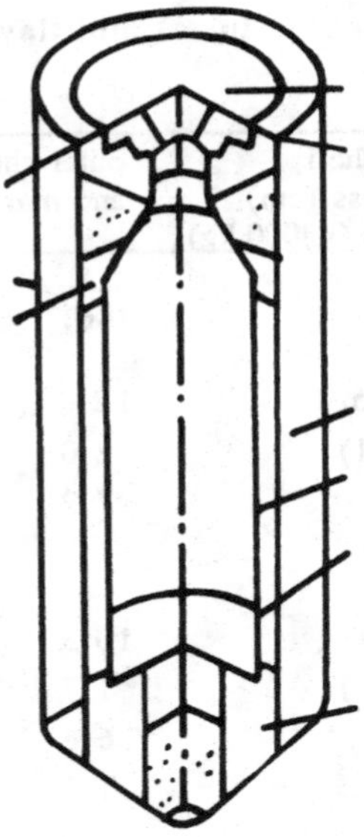

Figure 1. Model of hardened silo structure

Compressive strengths considerably in excess of 1000 atm can be obtained with the use of polymeric and silica additives, hardening under pressure at elevated temperatures, the reduction in matrix porosity, and the use of high strength, high alumina matrices in the concrete (13-15). The double wall silo structures discussed here are well suited for the application of warm curing under elevated pressures, allowing the ready use of these techniques.

It is evident that multi-wall steel and concrete structures are not directly suitable for application with static pressures on the order of 7000 atmospheres (see Chapter 12), since critical metal components and the concrete are likely to fail at static applied pressures on the order of 3000 atm. Such structures may well be able to survive considerably higher transient pressures, such as those of an adjacent nuclear explosion due to the short time enhancement in mechanical properties of the materials and energy absorption within the structure by deformation of the metal reinforcement and delamination of interfacial bonds.

Table 2. Relative induced stresses in a multilayer cylinder by an applied external pressure as a function of elastic ratio*

location	inner shell thick. (cm)	elastic ratio=3:1 outer shell thickness (cm)			elast ratio=2:1 outer shell thickness (cm)		
		2.5	5	10	2.5	5	10
σ(inner shell)	2.5	7.1	6.9	6.5	5.0	4.9	4.8
σ(cement composite)		2.3	1.3	2.1	2.5	2.4	2.3
σ(outer shell)		4.1	4.0	3.9	3.0	2.9	2.9
σ(eff. matrix stress)		1.1	1.0	.99	.78	.77	.74
σ(inner shell)	5	6.6	6.4	6.0	4.8	4.7	4.6
σ(cement composite)		2.1	2.1	2.0	2.3	2.3	2.2
σ(outer shell)		4.0	3.9	3.7	2.9	2.9	2.8
σ(eff. matrix stress)		.92	.89	.84	.70	.69	.67
σ(inner shell)	10	5.8	5.6	5.3	4.5	4.4	4.3
σ(cement composite)		1.8	1.8	1.7	2.1	2.1	2.09
σ(outer shell)		3.7	3.6	3.4	2..8	2.7	2.7
σ(eff. matrix stress)		.67	.65	.62	.57	.56	.54
σ(inner shell)	20	4.8	4.7	4.5	4.0	4.0	3.8
σ(cement composite)		1.5	1.4	1.4	1.8	1.8	1.7
σ(outer shell)		3.3	3.2	3.1	2.6	2.6	2.5
σ(eff. matrix stress)		.37	.36	.35	.39	.38	.37

* All stresses are calculated at the inner wall of the respective layer because this is the location of the peak stresses in the layer. A elastic modulus ratio of 3:1 can be obtained by an ~ 20% volume loading of steel wire in portland cement based concrete. A modulus ratio of 2:1 would require a ~ 40% loading. High strength concretes have higher elastic moduli and would require a reduced steel loading to possess these modulus ratios. The applied stress = 1.0. The inner radius is assumed to be 1 m and the outer radius 2 m.

Steel/cement composites

Rather than deploying the reinforcing steel in the form of separate shells, it can be distributed in the cement/concrete matrix in the form of reinforcing wire/bars/ribbon. While reinforcing steel bars may have tensile strengths on the order of 12,000 atm, low alloy reinforcing wire and amorphous iron/boron reinforcing ribbon have strengths on the order of 20,000 atm. By varying the placement of the reinforcement, a radial variation in the effective strength and elastic modulus of the composite may be obtained. A brief description of this approach is given in Appendix 2.

These composite structures do not appear to offer performance advantages over multiwall structures. They require the transportation of the full structure, a difficult job since the mass for a small silo can exceed 10^6 kg.

Silo and missile buffering from ground shock

The various structures considered above suggest that the engineering limits for silo resistance to static overpressure are approximately 3000 atmospheres for heavily reinforced cement/steel structures and 5000 atm for solid metal structures (while very thick walled metal structures of high strength steel could survive static pressures approaching 7000 atm, construction of such silos would be exceptionally expensive, requiring wall thicknesses of at least several meters of high alloy, high strength, high toughness steels). These numbers are less than the 7000 atm peak overpressure threshold for survival at the inner bowl cratering threshold (see Chapter 3), but the enhancement in silo survival due to the short time of the applied load suggests that silo survival at such loadings is feasible. Survival of the silo at applied overpressures in excess of this value, i.e. within the inner crater bowl, is not feasible. In the inner bowl region of the crater the rock is crushed and excavated from the ground, a process that takes hundreds of milliseconds. On such a time scale the mechanical properties of the silo materials approach static values. The large bending stresses and forces due to the flow of mobilized pulverized rock further aid in the destruction of the silo.

Hardening of silos for intense ground shock environments may require the deployment or modification of the material surrounding the silo to absorb energy from the pressure pulse and diffuse the blast pulse waveform. Such protective measures are particularly needed if the silo is attacked with earth penetrating warheads. Such warheads explode after penetrating a considerable distance into the ground, increasing the efficiency of energy coupling into ground shock by more than a factor of 10 over the ground shock coupling efficiency of an surface burst. It seems likely that such weapons would have a lower yield than an equal mass warheads designed for air burst, but the increase in efficiency in coupling the blast energy to the ground would allow such warheads to have larger destruction radii than the air burst weapon. Neither the US or the USSR currently have such weapons in their arsenals, but their development and deployment is under consideration.

Government statements (16,17) of the minimum targeting error for a groundburst 500 KT weapon suggest that this range is approximately 75±5 m. Comparison of this estimate with the inner bowl crater size in hard rock, 66 m, suggests that the inner bowl crater size can be used as a conservative measure of the destructive radius of a subsurface explosion. The crater radii in hard rock of weapons of various yields as a function of the depth of detonation (18) is plotted in Figure 2. The rapid increase in crater radius and hence silo destruction range for weapons of small to moderate yield as the warhead penetrates into the ground is obvious.In addition to increasing the destructive range of the warhead, subsurface detonation prevents the radiation burst from one warhead from destroying nearby warheads. Multiple warheads timed or fused for simultaneous subsurface detonation will set up coherent blast wave interactions. This would result in the formation of Mach stems and fronts within the ground. These effects would further increase the effective destructive range of the attacking warheads.

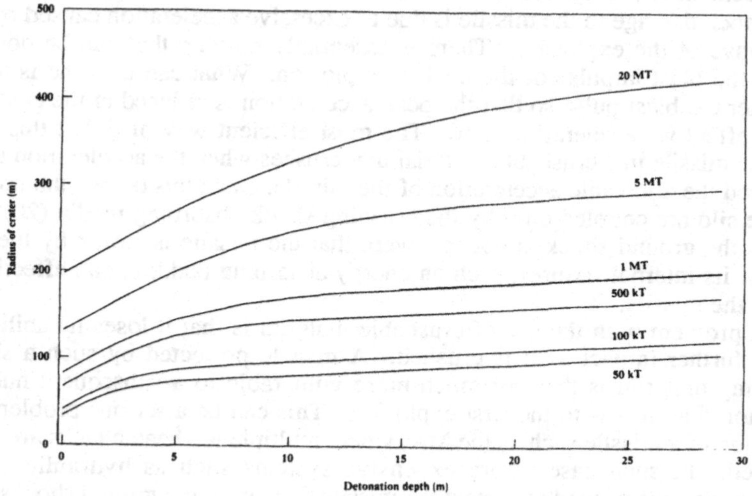

Figure 2. Crater radius as a function of yield and depth of detonation in hard rock.

I have no information upon the near field propagation of intense blast waves in hard rock or in highly dissipative materials. Glasstone's information upon the propagation of blast waves in deep water, an elastic material, does not provide information upon near source behavior and does not supply any functional information that would allow me to extrapolate the values needed from the data supplied. The government claims (19) that silos can be built to the inner bowl radius of a crater in hard rock suggest that such techniques can effect a factor of 2 to 3 decrease in peak load. This is confirmed by discussions with individuals familiar with the dynamic modeling of such interactions. Government presentations suggest several ways to enhance silo survivability in an intense ground shock environment. These are

•Surround the silo with a high strength crushable foam so that much time and energy is spent in crushing the foam and displacing the silo during the peak pressure period.

•Increase the energy absorption and variation in shock propagation velocities in the rock matrix. Drilling an array of holes into the rock surrounding the silo creates an array of stress concentrators in the rock around the silo. This problem was discussed in the earlier treatment of an isolated cylindrical cavity in a elastic medium subjected to an external pressure. The cavity induces a stress 2 times the applied stress in the cavity wall. The rock would fail under the blast wave, causing a scattering of the wave within the inhomogeneous medium, lengthening the blast time constant and reducing the peak load applied to the silo.

•Design the silo structure to be able to dissipate substantial amounts of energy within the silo structure itself. This will typically occur by deformation of steel members and reinforcement within the silo and debonding of fiber reinforcement within the matrix of the silo.

Shock damage to the missile is due to excessive acceleration caused by the blast wave of the explosion. There is essentially nothing that can be done to reduce the blast impulse of the nuclear explosion. What can be done is to try and filter the blast pulse so that the peak acceleration is reduced at the cost of a longer effective acceleration time. The most efficient way of doing this is to pack the missile in a crushable material that crushes when the acceleration starts to exceed the allowable acceleration of the missile. Once this occurs the missile and the silo are coupled only by the crushing shock absorbing media (20). As long as the ground shock is not so severe that the missile is struck by the silo walls or its internal fixtures, such an energy absorbing padding can effectively protect the missile.

A problem with the use of crushable material is that it loses its ability to absorb further impact as it is crushed. A missile protected by such a shock absorbing material is therefor much more vulnerable to a subsequent nuclear explosion than it was to the first explosion. This can be a serious problem for protection of a missile such as the MX, where multiple warhead attacks are to be expected. In such cases more expensive systems such as hydraulic shock absorbers may be needed to protect the missile from multiple ground shocks.

The critical factors in such shock isolation systems are the allowable acceleration of the missile, the corresponding required crushing strength of the energy absorbing media, and the available crushing distance provided before the mechanical load applied by the media and the silo wall exceeds the strength and acceleration limits of the missile. It is desired to have the crushable material absorb a given energy density, since too high an energy loading would cause failure of the material and too low an energy loading does not make efficient use of it. Since the cross-sectional density of the missile (parallel to its axis) (scales as its radius squared while its cross-sectional area (parallel to its axis scales as its radius, the necessary thickness of crushable material scales as the missile radius as well.

Conclusions

The resistance of missile silos to crushing by static applied overpressures appears to be limited to applied pressures on the order of 3000 atm. It is likely that such structures can survive peak airblast pressures on the order of 7000 atmospheres. Survival of silos when attacked with earth penetrating warheads appears to require the buffering of the silo from the ground and modification of the mechanical properties of the rock near the silo as well as substantial amounts of energy absorption and ductility in the silo. Hard silos with static overpressure limits >1000 atm appear to be easily made, and may be amenable to mass production techniques. Such silos, particularly for small missiles, could be made very inexpensively.

Appendix 1: A simple solution for the cylindrical shell

Consider an infinitesimally thin circular cylinder of radius r, under internal pressure p_i, and external pressure p_o. The cylinder will be taken to be long, so its length is >> r. Summation of the forces for the cylinder results in the following equation for equilibrium (21)

$$\sigma_t - \sigma_r - r\frac{d\sigma_r}{dr} = 0,$$

where is the radial stress. For simplicity consider the case where no axial load is applied along the cylinder. In this case the axial strain will be due only to the Poisson strain due to the radial and tangential stresses. Since the cylinder is long the axial strain will be a constant independent of radial position.

$$\epsilon_l = -\frac{\upsilon\sigma_r}{E} - \frac{\upsilon\sigma_t}{E}$$

This can be rewritten to yield

$$-\frac{E\epsilon_l}{\upsilon} = \sigma_r + \sigma_t = 2C_1$$

where E is the elastic modulus of the material, and ν is the Poisson ratio of the material. C_1 is a constant which will be determined later from the boundary conditions.

$$r\frac{d\sigma_r}{dr} + 2\sigma_r = 2C_1$$

$$\frac{d}{dr}(r^2\sigma_r) = 2rC_1$$

$$r^2\sigma_r = r^2C_1 + C_2$$

$$\sigma_r = C_1 + \frac{C_2}{r^2} \qquad \sigma_t = C_1 - \frac{C_2}{r^2}$$

Now the boundary conditions are that $\sigma_r = -p_i$ at the inner radius of the cylinder r=a, and that $\sigma_r = -p_o$ at the outer radius of the cylinder r=b. Solving for C_1 and C_2 and substituting into the above expressions we find

$$\sigma_{\theta\theta} = P_i\frac{(\frac{b}{r})^2+1}{(\frac{b}{a})^2-1} - P_0\frac{1+(\frac{a}{r})^2}{1-(\frac{a}{b})^2}, \quad \sigma_{rr} = -P_i\frac{(\frac{b}{r})^2-1}{(\frac{b}{a})^2-1} - P_0\frac{1-(\frac{a}{r})^2}{1-(\frac{a}{b})^2}$$

Appendix 2: Wire/concrete composite structures

The construction of anisotropic filimentary composite structures is a highly developed area of high performance structural design. Such structures are typically designed for maximum strength to weight ratios and use high performance materials such as fiberglass, graphite, or aramid fibers. The case of missile silos does not require low density, but it does require high strength and relatively low costs. In the discussions that follow the anisotropies that can be obtained by use of winding long filaments onto a preform will not be treated. The volume averaged strength and elastic modulus will be taken to be dependent upon the volume fraction of steel reinforcement in the concrete composite. Using such an approximation, the radial variation in the average stress may be determined (3).

$$\sigma_r = \frac{\lambda}{r^2}\int rE(r)dr \qquad \lambda E(r) = \sigma_t+\sigma_r$$

Let $E(r)=A+Br+Cr^2$, allowing a Taylor series approximation of the elastic modulus, and require that the applied pressure on the outer surface is b. The average radial stress is given by

$$\sigma_t = \frac{6A(1+(\frac{a}{r})^2)+4Br(2+(\frac{a}{r})^3)+3Cr^2(3+(\frac{a}{r})^4)}{6A(1-(\frac{a}{b})^2)+4Br(1-(\frac{a}{b})^3)+3Cr^2(1-(\frac{a}{b})^4)}\,P_0$$

and the average tangential stress is given by

$$\sigma_r = \frac{6A(1-(\frac{a}{r})^2)+4Br(1-(\frac{a}{r})^3)+3Cr^2(1-(\frac{a}{r})^4)}{6A(1-(\frac{a}{b})^2)+4Br(1-(\frac{a}{b})^3)+3Cr^2(1-(\frac{a}{b})^4)}\,P_0$$

Since the strains in the steel and the matrix must be the same (until the bond between the matrix and the steel reinforcement fails), the stresses in the cement and the steel are related by their respective elastic moduli. For common portland concrete and steel, this ratio is on the order of 1:6, respectively. High strength alloy steel reinforcing wire (~1 mm diameter) has a strength on the order of 20,000 atmospheres. For the composite to take full advantage of the strength of the wire it will need a strength on the order of 3000 atmospheres, a factor of 10 higher than the strength of portland concrete. Consider a two phase composite of materials α and β with elastic moduli E_α and E_β and volume fractions V_α and V_β.

$$E_{ave} = V_\alpha E_\alpha + V_\beta E_\beta = V_\alpha (E_\alpha - E_\beta) + E_\beta$$

The equality of the strains requires

$$\epsilon_{ave} = \epsilon_\alpha = \epsilon_\beta \rightarrow \frac{\sigma_{ave}}{E_{ave}} = \frac{\sigma_\alpha}{E_\alpha} = \frac{\sigma_\beta}{E_\beta}$$

yielding stresses

$$\sigma_\alpha = \frac{E_\alpha}{E_{ave}} \sigma_{ave} = \frac{E_\alpha}{A+Br+Cr^2} \sigma_{ave}, \quad \sigma_\beta = \frac{E_\beta}{E_{ave}} \sigma_{ave} = \frac{E_\beta}{A+Br+Cr^2} \sigma_{ave}$$

Table 3. Relative stresses in composite cylinders with linear radial variation in the elastic modulus

normalized elastic moduli		average stress		stress in steel		stress in concr		mass
inner wall	outer wall	inner wall	outer wall	inner wall	outer wall	inner wall	outer wall	
1	2	1.7	2.4	109.3	7.3	1.7	1.2	151
2	1	3.7	.85	11	5.1	1.8	.85	115
1	3	1.3	2.8	7.6	5.6	1.3	.93	302
2	3	2.1	2.1	6.3	4.3	1	.71	417
3	2	3.3	1.2	6.5	3.5	1.1	.59	381
3	1	4.2	.4	8.5	2.5	1.4	.41	230
2	2	2.7	1.7	8.0	5.0	1.3	.83	266
3	3	2.7	1.7	5.3	3.3	.88	.55	532
1	4	1.0	3.0	6.0	4.5	1.0	.75	453
4	1	4.6	.14	6.9	.85	1.1	.14	345
2	4	1.5	2.6	4.4	3.2	.73	.52	568
4	2	3.7	.85	5.5	2.2	.92	.42	469

* The elastic modulus of the matrix is taken to be 1 and the elastic modulus of the reinforcement is taken to be 6. The mass figures are for the mass of steel wire reinforcement required in a silo structure with a cylinder length of 15 m, an inner radius of 1 m, and an outer radius of 2 m. The mass of the hemispherical end caps are included.

Thus a steel and concrete matrix with a 20% volume fraction has a elastic modulus twice that of the concrete alone. The stresses in the concrete are reduced by a factor of 2 and those in steel increased by a factor of 3 with respect to the average stresses (22). Table 3 presents the relative stresses (with respect to the applied pressure) at the inner and outer surfaces of cylinders with linearly varying elastic moduli with specified relative elastic moduli at the respective surfaces. The associated stresses in the steel and concrete are also calculated

References and notes

1. S. Glasstone, Editor, *The Effects of Nuclear Weapons*, US Atomic Energy Commission, Washington, D.C. (1957).

2. Thin walled structures are structures with wall thicknesses much smaller than the radii of curvature of the walls. Typically thin walled structures are taken to have wall thicknesses < .1 times the local radius of curvature.

3. J. E. Shigley, *Mechanical Engineering Design*, McGraw-Hill, New York (1977(.

4. Della M. Roy, "New strong cement materials: chemically bonded ceramics," Science, Feb 1987, pp. 651-658.

5. Ramon L. Carrasquillo, "Production of high strength pastes, mortars, and concretes," Materials Research Society Proceedings: very high strength cement based materials, 1984, pp. 151-168.

6. Shuaib H. Ahmad and S. P. Shah, "Properties of high strength concrete concrete for structural design," ibid., pp. 167-181.

7. Antoine E. Naaman, "High strength fiber reinforced cement composites," ibid., pp. 217-229.

8. Sidney Diamond, "Very high strength cement-based materials-a prospective," ibid. pp. 233-243.

9 Farrohk F. Radjy and Kjell E. Loeland, "Microsilica concrete: a technological breakthrough commercialized," ibid., pp. 3096-312.

10. If all three principle stresses are compressive, failure will occur at a higher level than would otherwise be expected. Materials such as concrete do not typically fail under hydrostatic loads. The stress associated with the hydrostatic component may be subtracted from the stresses before determination of the effective stresses for failure determination. In the case of a deformable material a similar procedure is followed in which those components of the stress field that are not deforming the material are subtracted from the stress field before determination of the failure threshold.

11. Assuming that the joint of the silo top with the cylindrical section and the associated top clearance machinery is not more vulnerable than the structure of the silo.

12. David R. Lankard, "Slurry infiltrated fiber concrete (SIFCON): properties and applications," Materials Research Society Proceedings: very high strength cement based materials, 1984, pp. 277-286.

13. W. Sinclair and G. W. Groves, "High strength cement pasts," Journal of Materials Science, Volume 209, 1985, pp. 2846-2852.

14. R. Morayan Swamyl, "Polymer reinforcement of cement systems, Part 1: polymer impregnated concrete," ibid., Volume 14, 1979, pp. 1521-1553.

15. N. B. Eden and J. E. Bailey, "On the factors affecting the strength of portland cement," ibid., Volume 19, 1984, pp. 150-158.

16. Quotation from Air Force Secretary Cooper, "SS-18 not capable of 2509 foot CEP," Defense Daily, 22 May 1985, p. 121.

17. George Ullrich of the Shock and Strategic Structures Division, Defense Nuclear Agency, quoted in Jonathan Medalia, *Single Warhead ICBMs*, Congressional Research Report #83-1096F, Washington, D.C. (1983).

18. Calculated using an unclassified code from the Defense Nuclear Agency.

19. M. I. Kovel, "DNA ICBM technical R&D program," Shock and Vibration Bulletin, Volume 54, 1984, pp. 23-41.

20. J. C. Snowdon, "Isolation of impact transients," presented in a vibration workshop at the Pennsylvania State University, University Park, PA, Nov 1974.

21. In elastic mechanics tensile stresses are defined to be positive and compressive stresses are defined to be negative. This is opposite the standard convention for hydrostatics, but is the accepted standard.

22. Assuming that the mixture has no preferred orientation, which would result in different load sharing results.

Part V Chapter 10

Engineering of mobile and other deceptive deployments

John Michener

Introduction

Increases in ICBM accuracy are rendering even the hardest silos potentially vulnerable to ICBM attack and destruction by single warheads. Deployment modes in which the location of the missile is uncertain (from the attackers perspective) are of potential interest to reduce the likelihood or increase the cost of a preemptive attack. The megatonnage necessary to destroy such dispersed deployment schemes is dependent upon the area available for dispersal and the hardness of the missiles and their protection from and resistance to the weapon effects. The cost effectiveness of these deployment schemes is determined by the relative costs to build them with respect to the cost to destroy them.

Overpressure blast protection

The blast overpressure P (in atm) from a ground burst of a nuclear weapon of yield Y (in megatons, MT) as a function of radius r (in km) may be approximated by (see Chapter 3)

$$P = 6.31\ Y/r^3 + 2.20(Y/r^3)^{1/2}$$

At peak overpressures < 10 atm the radius of destruction of a nuclear weapon is increased by detonating the weapon above the target point. The maximum destruction radius as a function of yield and overpressure is plotted in Figure 1 (1).

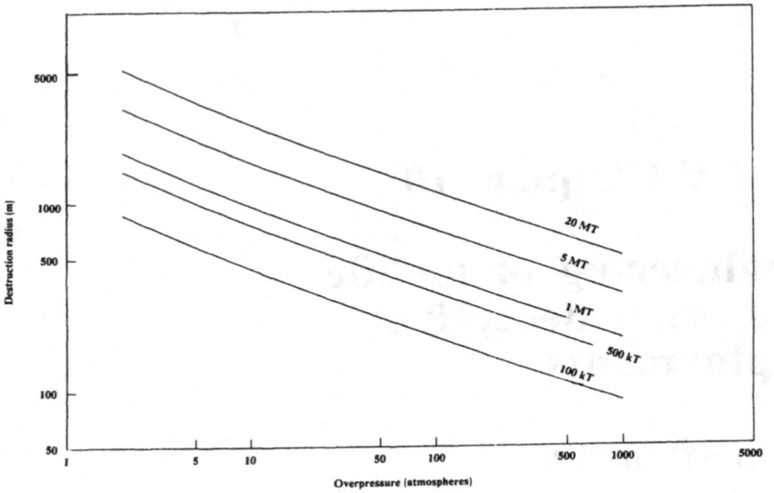

Figure 1. Destruction radius as a function of yield and overpressure.

If an hexagonal close packed barrage pattern (with the destruction radii just contacting one another is used, approximately 90% of the total area is included within the circles of destruction (2) . Hence the area that can be barraged by N warheads (ignoring targeting errors) is given by $A = 1.1\pi r_d^2 N$ where r_d is the destruction radius of the warhead. Most modern missiles have high throwweights and multiple warheads. The yield of the warheads as a function of the number of warheads carried may be approximated by (3) $Y=Y_o/f^{1.5}$, where f is the number of warheads carried by one missile and Y_o is the yield that the missile would carry if it were a single-warhead missile. Consequently, the number of warheads carried by a missile as a function of the yield of the warheads and the payload capability of the missile can be approximated by $f = (Y/Y_o)^{2/3}$. Since the area within the destruction radius scales largely as $Y^{2/3}$, the area that can be barraged to a given overpressure is largely independent of the missile payload fractionation.

Radiation protection

The gamma rays and neutrons released in the explosion and initial decay of radioactive by-products carry a large amount of energy and can do substantial damage to considerable radii. The radiation dose is dependent upon atmospheric conditions since the atmosphere absorbs the radiation, with the absorption length for a reference atmosphere being ~ 300 m. The radiation yield of neutrons and gamma rays is dependent upon the fission fraction (the fraction of the energy output of the device due to fission) of the weapon. This is usually about 0.5. Atmospheric absorption is dependent upon atmospheric density, which is affected by the yield of the weapon. The detonation of high yield devices significantly reduces the atmospheric density and hence the atmospheric shielding of gamma rays from short lived fission products. The curves of Figure 2 (4) show the total prompt tissue dose (5) for groundburst explosions of various yields as a function of the peak static overpressure. It is evident that against relatively soft targets small yield weapons are much more efficient radiation weapons than are large yield weapons.

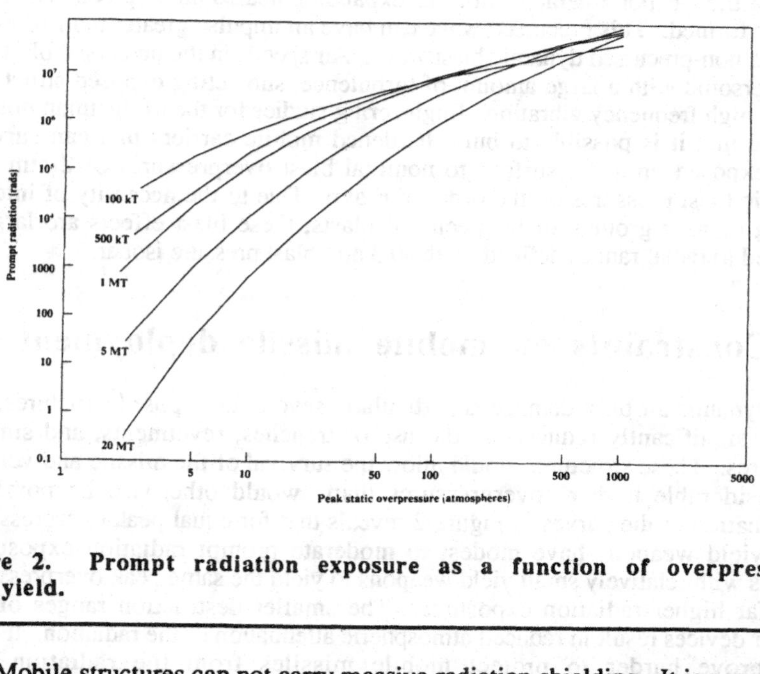

Figure 2. Prompt radiation exposure as a function of overpressure and yield.

Mobile structures can not carry massive radiation shielding. It is necessary to determine the relative hardness of mobile systems to both radiation and blast effects of the attacking nuclear weapons. Typical memory elements, processors, and application specific integrated circuits have radiation hardnesses ranging from 10^4 to in excess of 10^6 rad. People are far more vulnerable: The lethal dose is about 500 rad and doses in excess of several thousand rads rapidly disable exposed individuals. Published pictures of the Midgetman mobile vehicle show the presence of radiation shielding on the vehicle. This vehicle is designed to survive a nominal 2 atm overpressure from a 1 MT yield airburst. Such a explosion would expose an unshielded individual to roughly a 700 rad prompt tissue dose. If the barrage used 100 kiloton warheads the prompt radiation exposure would be roughly 20,000 rad.

Dynamic air blast damage

Exposed structures, such as exposed mobile missile carriers, are vulnerable to damage by the dynamic air blast and the high velocity debris that is mobilized by the dynamic air blast. The impulse of the dynamic blast can easily turnover, displace, or pickup exposed objects. The problem of dynamic blast is particularly severe if the weapon is detonated over dark ground in clear weather, where the thermal radiation from the fireball can heat the air adjacent to the ground. Under such conditions the near ground air can be heated to several thousands of Kelvins, significantly reducing its density. When the primary blast wave of the weapon interacts with this expanding heated air a "precursed" blast wave is formed. This precursed wave can have an impulse greater than 10 times that of a non-precursed dynamic blast wave. Air speeds in the precursed blast can be supersonic with a large amount of turbulence, subjecting exposed structures intense high frequency vibration. Engineering studies for the Midgetman missile indicate that it is possible to build hardened mobile carriers that can survive when exposed on a flat surface to nominal blast overpressures of 2 atm and dynamic blast pressures on the order of 6 atm. Due to the necessity of intense heating of near ground air in precursed blasts, these blast effects are largely confined to radial ranges defined by the 0.3 atm blast pressure isobar.

Constraints on mobile missile deployment

Dynamic air blast damage is particularly severe for exposed structures, but can be significantly reduced by the use of trenches, revetments, and similar structures. These structures would allow the survival of the missile and vehicle to considerable higher overpressures than would otherwise be possible. Examination of the curves in Figure 2 reveals that for equal peak overpressures large yield weapons have modest to moderate prompt radiation exposures. Attacks with relatively small yield weapons to yield the same peak overpressures have far higher radiation exposures. The smaller destruction ranges of the smaller devices result in reduced atmospheric attenuation of the radiation. It may well prove harder to protect mobile missiles from the radiation and electromagnetic effects of nuclear detonations than from the mechanical damage mechanisms if small yield weapons are used in the barrage.

Mobile deployment options

There are two mobile missile systems currently under consideration in the US. One of these is the deployment of small single warhead missiles in armored ground carriers that are hardened to survive the blast, radiation, drag, and debris effects associated with a 2 atmosphere overpressure. These carriers would probably be stored at ICBM silo bases and deployed into the surrounding countryside at times of risk or attack warning (6). The other mobile system under consideration is the deployment of 10-warhead MX missiles on armored heavy trains (7). These trains would be kept in garrisons and would be distributed onto the US rail network at times of risk or strategic warning. Such a rail system relies upon strategic warning systems to provide from 4 to 24 hours of warning for dispersal of the trains onto the rail network.

Such trains are relatively soft targets. A train on an open track can be damaged by a blast overpressure of about 0.2 atm perpendicular to the train or about 0.7 atm along the orientation of the train. At these low damage thresholds precursed blast effects and radiation are not major impediments to the system survivability. The orientation dependence of the blast damage threshold results in an elliptical destruction footprint of the attacking nuclear weapon.

These deployment options are discussed in detail in Chapters 7 and 8.

Mobile carriers can be protected from blast effects far more easily than they can from the radiation effects. Since mobile carriers can not carry enough radiation shielding to allow crew survival at design overpressures in excess of 2 atm, any manned mobile system that is expected to survive higher overpressures must provide the radiation protection in a massive fixed protective structure. This would typically involve several m of earth and/or reinforced concrete.

Lateral semi-silos

Two deployment options for mobile missiles will be further considered in this paper, both of which offer adequate radiation for the missiles and crews. Even light construction earth sheltered structures offer substantial protection against blast and radiation (8). . The construction of stronger structures to survive overpressures in the 100 atm range is not difficult. The first of these deployment options for mobile missiles is suitable only for the Midgetman missile due to its requirement for small size vehicles. In this mode the missile would be deployed on a vehicle with minimal diameter which would carry a spherical shell plug at the end of the vehicle. The vehicle would be designed to mesh into a lateral semi-silo: a high strength concrete and steel structure/silo placed laterally on the ground and covered (except for the entrance) with earth. The carrier would be driven into the lateral semi-silo so that the spherical shell cap seated into the reinforced walls of the structure. Figure (3) illustrates this deployment mode, which does not require that the fixed installation have any machinery, since the carrier is used as the door and the door opener.

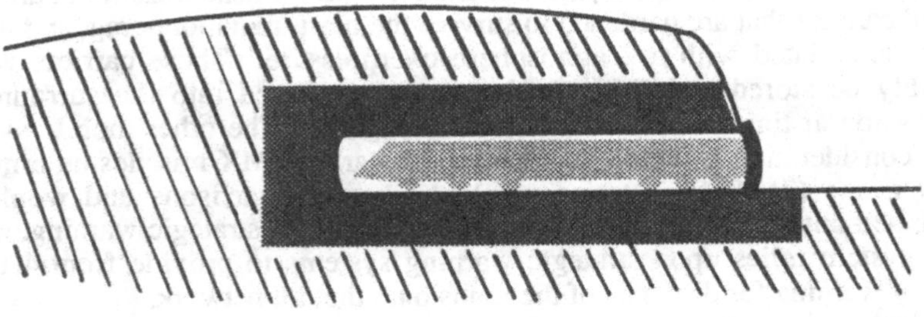

Figure 3. Lateral semi-silo bunkers.

A spherical section end cap with a radius of 10 m and a thickness of 10 cm would have an induced stress of 5000 atm when exposed to a 100 atm overpressure. A steel cap with a cross-sectional area of 3 m by 4 m would have a mass of some 10 tons, not an excessive value. A carrier with a smaller cross-section would require a less massive plug. If the tunnel had an inner radius of 2.5 m and an outer radius of 4 m, the inner wall stress when exposed to 100 atm would be roughly 330 atm, not an unreasonable value for a conventional reinforced concrete structure. Heavy reinforcement would be needed where the end cap is seated to distribute the high contact loads. It appears that lateral semi-silos may provide a means of protecting mobile missiles to overpressures approaching 100 atm. The cap would provide modest radiation shielding capabilities (radiation reduction factors of 10 to 100), depending upon its mass and construction. The missile electronics would be further shielded by the length of solid propellant between the door and the electronics. The cap would not provide adequate shielding for the crew. It would be necessary to build a small tunnel/shelter below the main bay for the crew to shelter in during the attack. Such a construction would allow several meters of earth and concrete for radiation shielding.

Low cost silos in a multiple silo array

The other deployment option to be considered here is to deploy the missiles in a "multiple silo" field of much more numerous low cost silos. Silo based missile deployment is well understood, and requires individual targeting of specified silos. In Chapter 9, the design of multiwall cylinders for high strength missile silos has been discussed. Such structures are suitable for the mass production of small silos since the inner and outer shells could be produced as single pieces at a factory and moved to location where they would be inserted into the ground and the loaded with high strength concrete. The masses of the steel/iron shells given in Table 1 of Chapter 9 are not excessively high for readily available carriers or the road network, allowing the transport of the shells from a central manufacturing location to the installation site.

Because of the geometric scaling of the mass of missile silos, the mass of the shells necessary for the construction of silos to protect missiles such as the MX would be far greater than the mass of the shells necessary to protect smaller missiles such as the Midgetman. Mass production and transportation of single piece shells for MX sized missile silos is probably not feasible. While survival of such silos requires sophisticated materials processing if the structures are to survive overpressures on the order of 3000 atm, much cheaper materials may be used at lower overpressures. High strength nodular cast iron has enough strength and ductility to be of potential interest for this application (roughly 6000 atm compressive failure strength) as well as its associated easy castability, allowing relatively easy manufacture of the iron shells. In combination with wire/ribbon reinforced high strength concrete, such iron or steel shells should allow the inexpensive construction of silos able to survive 1000 atm overpressures.

Such silos are not vulnerable to area barrage attacks and would have to be individually targeted. Since the attacker could not know which silo contained a missile, all silos would have to be attacked - an unfavorable exchange for the attacker if the silos are relatively inexpensive and much more numerous than the missiles. The silos would have to have the cap and cap opening mechanisms because their mass is excessive for the carrier itself. The missiles could be distributed among a much more numerous field of silos, which would be partitioned into sections. At specified times all silos in a specified sections could be opened to allow counting of the number of missiles in the section. The missiles would then be reemplaced in other silos using movable covers and equipment so that the location of the missiles in the silos would once again be unknown. A schematic for this dispersal mode is shown in Figure 4.

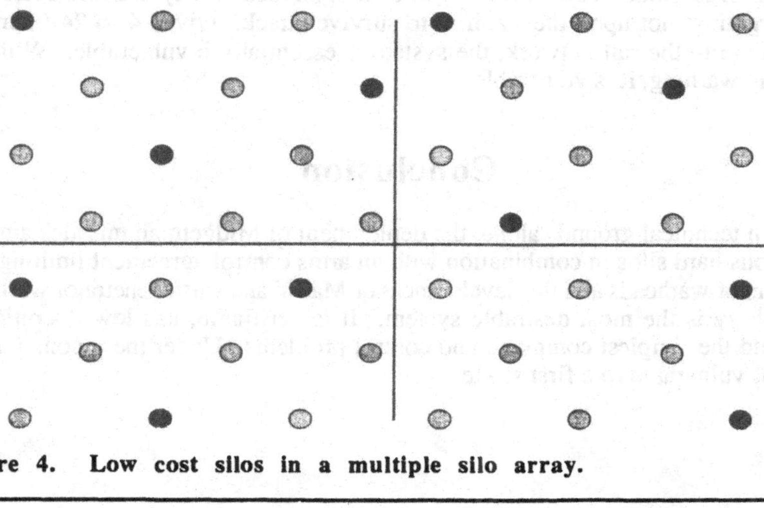

Figure 4. Low cost silos in a multiple silo array.

Rail mobile deployment options

Unlike road/ground mobile systems, rail mobile systems can carry substantial amounts of armor and radiation shielding (9). Rail mobile carriers would be heavily armored to protect the systems from vandals, terrorists, and saboteurs. As discussed previously, heavy railroad vehicles are damaged or displaced from the track at blast overpressures of from 0.2 to 0.7 atm. Because of this, rail mobile deployment modes are far move vulnerable to blast overpressure than hardened mobile launchers.

The rail mobile system could be very vulnerable to a "bolt from the blue" attack. If the trains were deployed in clusters of relatively soft bunkers at a few land bases the system could be destroyed by a few, inaccurate missiles such as SLBMs. This would result in severe pressure to launch these missiles upon warning, since they may be exceptionally vulnerable to a first strike. This problem can be essentially alleviated if the trains are dispersed onto numerous bases in hard, buried, bunkers hardened to some 100 atm. By dispersing the bases adequately it would not be possible to attack the system with MIRVed SLBMs since the bases would be too dispersed to allow any missile to attack two trains. This would raise the cost to attack the 25 trains to 50 SLBM missiles, a substantial attack (equivalent to some 350 half-megaton weapons) that would have substantial collateral effects upon the country.

While relatively low cost means of hardening such rail mobile systems are available, they tend to create vulnerabilities with targeting of the train from space or potential problems with arms control verification of the numbers of rail mobile systems deployed. Because of the susceptibility of a rail mobile system to targeted attack, hardening of the train or its surroundings against nuclear weapon effects does not appear to be justified. Any hardening effort that renders the train distinguishable is counterproductive. The rail mobile trains are far more valuable targets than individual Midgetman launch vehicles since each train carries 2 or 4 MX missiles (with 20 or 40 high accuracy warheads). The survival of rail mobile systems is dependent upon their ability to avoid detection and targeting, not upon their ability to survive attack. Given 4 to 24 hours to disperse onto the rail network, the system is essentially invulnerable. Without strategic warning, it is vulnerable.

Conclusion

On technical grounds alone, the deployment of Midgetman missiles among numerous hard silos in combination with an arms control agreement limiting the numbers of warheads and the development of MaRV and earth penetrator warhead technology is the most desirable system. It is verifiable, has low operational cost, and the simplest command and control problems. Under these constraints, it is not vulnerable to a first strike.

Reference and notes

1. Calculated using an unclassified code from the Defense Nuclear Agency.

2. By using an additional barrage, also a hexagonal close packed array, targeted at the interstices missed by the first barrage, a 100% coverage can be obtained as well as substantial redundancy against occasional missile and warhead failures.

3. I. Bellamy, *Nuclear Vulnerability Handbook*, Center for the Study of Arms Control, Lancaster, England (1981).

4. Calculated using an unclassified code from the Defense Nuclear Agency.

5. Radiation from the decay of short lived fission products makes a sizable contribution to the radiation yield of the weapon. This source is hard to calculate due to proximity of the target to the fireball for peak overpressures > 5 atm. This source is not included in these curves, which are therefore conservative estimates of the radiation yield of the weapon.

6. If deployed upon strategic warning, as in the case of rail mobile systems, such carriers would be essentially impossible to target or verify. If caught in garrison, this mode could be exceptionally vulnerable to SLBM attack.

7. The vehicles necessary to carry 4 missiles, their launch control car, and the associated support cars for security require a short train, with a minimal length of several hundred meters.

8. Glasstone reports that corrugated steel Quonset huts (10 gage steel, 7 m span) covered by some 1 m of soil but otherwise exposed at the surface require in excess of 2 atm overpressure to crush. Burial of such a hut so that its top is below the surface, protecting it from the dynamic pressure, required 3 atm overpressure to crush. Heavier structures without the aerodynamic exposure of such structures, and with far thicker coverage would be far more resistant.

9. Rail cars with masses of 500 tons are not uncommon in the transportation network and do not place exceptional loads upon the rail system.

Part V Chapter 11

Survivability of superhard silos

Art Hobson

Introduction

Advances during the late 1970s in silo strengthening techniques, in strength testing, and in knowledge of primary nuclear weapons effects, have revolutionized missile silo construction. Whereas the ultimate compressive strength of solid concrete (200 atm, 3000 psi) had been assumed to be the limiting strength for practical silos, it is now known that structures 10 or 20 times stronger are possible. The key to the new "superhardening" is the use of steel not only as reinforcing bars as in older silos, but also as thick cylindrical shells of steel lining the inner and outer perimeters of the silo. In addition, new silo designs make use of stronger missile components, thicker silo walls, a protective canister to hold the missile, and a "rattlespace" between the inner steel liner and the canister containing shock-absorbing materials (1-6). In Chapter 9, John Michener describes the engineering design and characteristics of superhard silos. Here, I analyze the survivability of an ICBM force, either Midgetman or MX, housed in such silos.

Silo strengths of at least 3500 atm (50,000 psi) are certainly achievable with the new techniques, and 7000 atm or more is probably achievable (1,3,5-7). These predictions, based on high-explosive tests with scale models, should be compared with the strength of present Minuteman silos, about 200 atm.

The meaning of "superhard"

For *conventional silos* such as Minuteman silos, the primary damage mechanism is airblast pressure. But superhardening can make silos so resistant to airblast that they will survive anywhere outside the crater (5,8-11). There is general agreement that no silo can survive inside a nuclear crater, regardless of the silo's strength against airblast. Within the crater, the silo is destroyed not by airblast but rather by cracking, deformation, and uprooting, due to ground motion during the formation of the crater.

Figure 1 shows the precise meaning of survival "outside the crater". A nuclear crater has three distinct parts: an inner bowl that is dug out and depressed by direct ground shock, an outer bowl that is depressed downward by airblast from above, and an above-ground debris pile. According to the Defense Nuclear Agency, silos can survive outside the inner bowl, but not inside (5,11).

By *superhard silos*, we will mean silos that are survivable at the edge of the crater's inner bowl but not inside. Superhard silos must withstand not only the airblast at the edge of the inner crater but also the ground compression of the outer bowl and the accompanying silo dislocation.

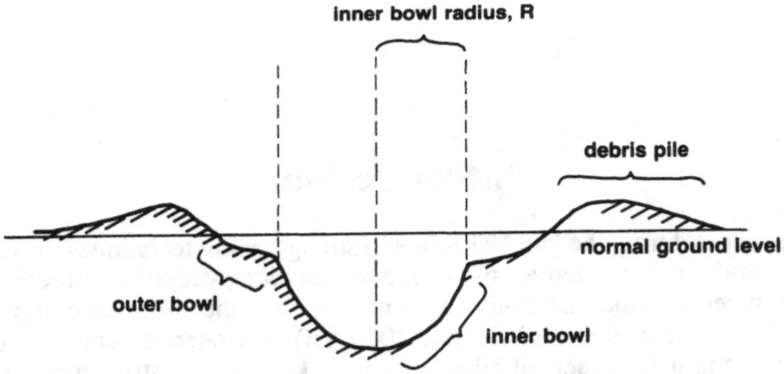

Figure 1. Cross-section of a nuclear crater.

Calculation of superhard silo strength (12)

Although figures are not available in the open literature, we can use the theory of silo destruction to estimate the strength of a superhard silo. Since the inner bowl's volume should be roughly proportional to the energy released,

$$R^3 \approx c\,y \tag{1}$$

where R is the radius of the inner bowl (Figure 1) and y is the yield. The proportionality constant c will depend on the geology surrounding the silo, being smaller for harder geologies. According to Chapter 3 Eq. (4), the airblast pressure at distance r from a surface burst is, for pressure greater than 100 atmosphere, $P = 7.04\ y/r^3$. At the edge of the inner bowl, this becomes $P_o = 7.04\ y/R^3$ or, using Eq. (1),

$$P_o \approx 7.04/c \tag{2}$$

As expected physically, *the critical pressure is independent of yield* and dependent only on the surrounding geology. By definition, a superhard silo must withstand at least this much airblast pressure. Furthermore, it would be superfluous for a superhard silo to withstand much more than this "critical pressure" P_o, for higher pressures would be felt only inside the inner bowl where the silo would be destroyed by other effects anyway.

To evaluate c, we need the crater size of at least one actual nuclear blast, in the geology likely to be chosen for superhard silo emplacement. It is possible to estimate the relevant data from the open literature. The standard reference (13) gives R = 120 m at y = 0.5 MT in supersaturated coral, which is much softer than likely US superhard silo basing areas. A Defense Nuclear Agency official stated in 1983 that 0.5 MT produces a 60-90 m crater (14). Another DNA official recently confirmed the 60-90 m range, assuming a "beneficial" (i.e. hard) geology (15). The most specific figure was given by an Air Force Assistant Secretary who stated, in 1985, that an 0.5 MT blast produces a crater of radius 75 m in typical superhard silo geology (16). It seems reasonable to conclude that R is about 75 m, say 75±5 m. Putting this into Eq. (1), we find

$$c = (8.5 \pm 1.5) \times 10^{-4}\ \mathrm{km^3/MT}, \tag{3}$$

and Eq. (2) implies

$$P_o = 8500 \pm 1500\ \mathrm{atm}. \tag{4}$$

This is the airblast pressure at the edge of the inner bowl of any nuclear crater in favorable geology, regardless of yield. It is the airblast pressure accompanying the ground shock that is needed to dislodge the last cubic meter of earth at the edge of the crater. Superhard silos must be able to withstand this pressure. As a check on our calculations, our calculated result, 8500±1500 atm, is consistent with the roughly 7000 atm thought to be achievable (1,3,5-7).

Survival probabilities for superhard silos

The most lethal Soviet missile, the SS-18, carries an 0.5 MT warhead with a CEP of 250-300 m (Chapter 3). This warhead cannot be very lethal against superhard silos, because its CEP is so much larger than its 70-80 m crater radius. On the other hand, the US MX warhead, as well as the planned Midgetman and Trident II warheads, all carry yields of about 0.5 MT with CEPs on the order of 100 m, so these have a much larger probability of destroying superhard silos.

Superhard silo survivability, for an arbitrary attacking warhead of lethality L, can be evaluated as follows: For superhard silos, by definition, the radius of destruction RD equals the crater's radius: RD = R. Thus Eq. (3) of Chapter 3 implies

$$SSPS = 0.5^x \quad \text{where } x = R^2/CEP^2 \qquad (5)$$

From Eq. (1), $R^2 = (cy)^{2/3}$. Inserting the mid-range value of c from Eq. (3), $R^2 = 0.009\ y^{2/3}$. Thus from Eq. (5) we get

$$SSPS = 0.5^x \quad \text{where } x = 0.009\ y^{2/3}/CEP^2,$$

from which

$$SSPS = 0.5^{L/110} \qquad (6)$$

where $L = y^{2/3}/CEP^2$ is the attacking warhead's lethality in units of $MT^{2/3}/km^2$. Equation (6), giving the survival probability of any superhard silo, is graphed in Figure 2. Note that attacking warheads need a lethality of about 110 or more to have a better than 50% chance of destroying superhard silos.

The SS-18 mod 4's lethality (per warhead) is 8.5±1.5 (Chapter 3), from which Eq. (6) gives SSPS = 94-96%, a high survival rate. But the Soviets also have an older *single* warhead "mod 3" version of the SS-18, carrying a large 20 MT yield with the same 250-300 m CEP as the mod 4 (17). In response to US superhard silos, the Soviet could re-deploy these. This warhead's lethality is 100±20, implying SSPS = 47-60%. Thus superhard silos might have survival rates rates of only 47% today. This survival rate could not be decreased by double-targeting because of fratricide: Both warheads would need to be groundburst to produce the ground shock to destroy superhard silos, and these warheads would explode within a distance of roughly the crater radius from each other. Double-targeting could, however, be employed to compensate for duds.

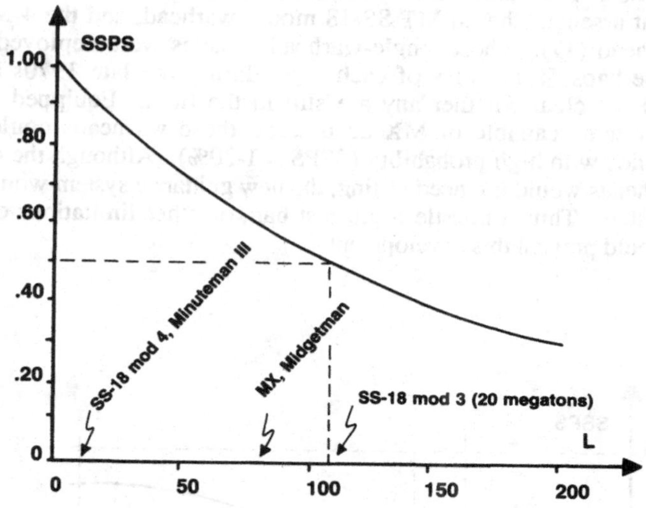

Figure 2. Single shot probability of survival (SSPS) of a superhard silo, as a function of the attacking warhead's lethality L, Eq. (6). For comparison, the lethalities of several warheads are indicated.

More generally, Figure 3 graphs the superhard silo survival probability calculated from Eq. (6), as a function of the CEP of the attacking warhead, for three different attacking warhead yields: 20 MT, 4 MT, and 0.5 MT. As we have seen, 20 MT is representative of single-warhead heavy ICBMs such as the SS-18 mod 3. A yield of 0.5 MT is representative of today's highly MIRVed ICBMs such as the SS-18 mod 4 or the MX, and would also be the yield of a typical SLBM such as the SS-N-20 or SS-N-23 if it were 3-MIRVed. An intermediate yield of 4 MT is also graphed; this is the yield of the single-warhead SS-19 mod 2, and is a plausible yield for a 3-MIRVed heavy ICBM or for a single-warhead SLBM.

Figure 3 shows that superhard silos will be vulnerable to inertially guided missiles of sufficiently high accuracy and yield, but that the demands on the attacking missiles are severe. The absolute accuracy limit for inertial guidance is around 75 m, because re-entry errors, due to buffeting in the atmosphere, are at least this large (18). The practical limit is probably the 100 m CEP attained by the MX (19). This means that none of the missiles deployed or tested in the past decade (SAS-18 mod 4, SS-19 mod 3, SS-24, SS-25, Minuteman III, MX, Trident I, Trident II) will be able to destroy superhard silos, even if their guidance systems rival the MX's accuracy, because their yields are all 0.5 MT or less. Figure 3 shows that superhard silos are not very vulnerable (SSPS ≈70%) to such missiles.

Even with MX accuracies, large 4-20 MT warheads are needed to destroy superhard silos. The Soviets have at least two plausible warhead candidates in their present arsenal: the 20 MT SS-18 mod 3 warhead, and the 4 MT SS-19 mod 2 warhead (17). These single-warhead missiles were deployed in small numbers, perhaps 50 missiles of each type, during the late 1970s and early 1980s; it is not clear whether any are still in the field. Equipped with new guidance systems capable of MX accuracies, these warheads could destroy superhard silos with high probability (SSPS = 1-20%). Although the 4 MT and 20 MT warheads would not need testing, the new guidance system would need to be flight tested. Thus a missile flight-test ban, or other limitations on missile accuracy, could prevent this development.

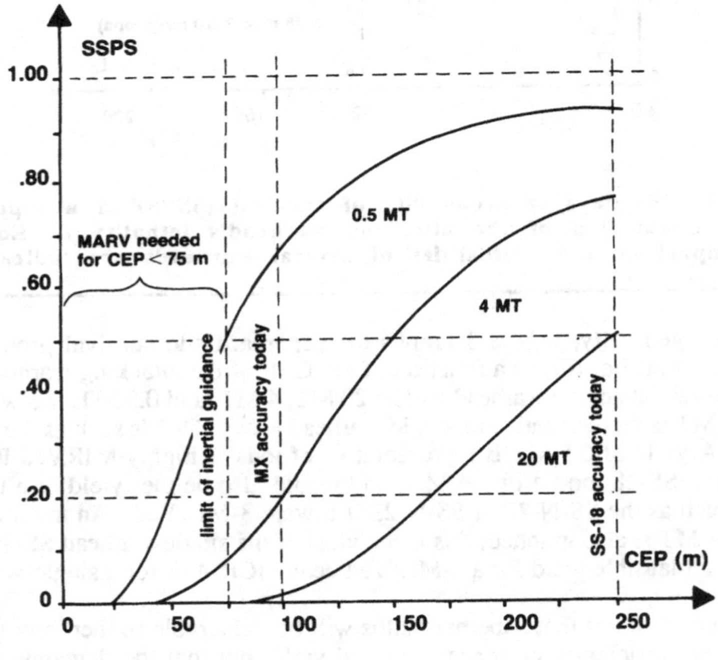

Figure 3. Survivability (SSPS) of a superhard silo attacked by warheads of 3 different yields, as a function of inaccuracy (CEP). The attacking warhead must be a MaRV (i.e. maneuverable during reentry) to attain a CEP less than 75 m. A yield of 0.5 MT is typical of today's highly MIRVed medium or heavy ICBMs, or of a hypothetical 3-MIRVed SLBM. 4 MT would be typical of a 3-MIRVed heavy ICBM or of a medium single-warhead ICBM or SLBM. 20 MT is the yield of the single-warhead heavy ICBM SS-18 mod 3.

This high-yield attack against superhard silos would have to be very large. The Soviets would need to use one missile against each superhard silo, and against say 500 Midgetman silos this would consume most of the 308 SS-18s and 330 SS-19s in today's inventory. The total yield of such an attack would be enormous, at least 2000 MT and perhaps much more. The fallout and atmospheric dust from these groundbursts would also be enormous.

A more plausible development might be Soviet deployment of a new MIRVed ICBM carrying 4 MT individual warheads. A heavy missile such as the SS-18 should be able to carry 3 such warheads. It is possible that, if the warheads were sufficiently compact, three of them could be carried by a medium-sized missile such as the SS-24. Assuming MX accuracies, only 170 such 3-MIRVed missiles would be needed to attack 500 superhard Midgetman silos with a high destruction rate (SSPS ≈20%). Development of the warhead would require testing of at least the smaller "primary" nuclear device that triggers the larger thermonuclear device, and the new MIRVing would require flight testing. Thus a low-level nuclear test ban or a missile flight test ban could prevent this development.

Another development may be high-accuracy SLBMs capable of destroying superhard silos. A typical recent Soviet SLBM, the SS-N-20 or SS-N-23, would carry about 4 MT if it were un-MIRVed. With MX (or Trident II) accuracy, such a missile could destroy superhardened silos with high probability (SSPS ≈20%). This SLBM threat is much less plausible than the ICBM threat, both because it is far from certain that the Soviets will develop such high-accuracy SLBMs, and because 500 SLBMs (over 30 submarines) would be needed to attack 500 Midgetman silos.

Superhard silos have been suggested not only for Midgetman, but also for MX basing. Because of the high exchange ratio (warheads destroyed/warheads used), an attack on superhard MXs is much more plausible than one on superhard Midgetmen. For example, 50 large and accurate single-warhead missiles could attack 50 MX silos holding 500 warheads. The Soviets have deployed on the order of 50 single-warhead SS-18s, and 50 single-warhead SS-19s, during the past decade. Re-deployment of these, with new MX-type guidance systems, would make a superhard MX force highly vulnerable. So superhard MXs make less sense than superhard Midgetmen.

MaRVs and earth penetrators

Superhard silo could be extremely vulnerable to either guided or earth-penetrating warheads. Guided warheads could decrease the CEP far below the 100 m crater radius of a surface-burst half-megaton warhead, while earth penetrators could produce much larger craters for a given yield.

Guided reentry into the atmosphere requires a homing device directing the warhead toward the target during reentry, and a mechanism for maneuvering in the atmosphere. Such maneuvering reentry vehicles are known as MaRVs (20). The United States has been developing and flight testing MaRVs for over a decade, and it is thought that a terminally guided MaRV could be ready for deployment on either the MX or Trident II during the 1990s. The US Pershing

II intermediate range ballistic missile, deployed in Europe but being removed under the INF agreement, already carries a guided with a CEP of 20-45 m (21). The guidance and maneuvering systems make the Pershing II warhead much heavier than other warheads of similar yield. Terminally guided ICBMs and SLBMs have not yet reached deployment stage because, for one thing, their longer range make the weight penalty more of a problem than for the Pershing II. The Air Force is currently planning a five year development program for intercontinental MaRVs, perhaps in order to be able to target Soviet mobile ICBMs (20). As long ago as 1960, the Soviets began to develop a terminally guided intermediate range (650 km) ballistic missile to be launched from submarines, but technical difficulties prevented its deployment (22).

Figure 3 shows that guided warheads of 50 m accuracy would make superhard silos highly vulnerable to half-megaton warheads. Terminal guidance makes such CEPs plausible. This development would make superhard silos vulnerable to a small number of highly MIRVed ICBMs or SLBMs---precisely the situation that Minuteman-class silos are in today. However, this development is strongly dependent on developing lightweight MaRVs, because each of the MIRVed warheads must carry a separate guidance and maneuvering system.

Earth-penetrating warheads explode after penetrating some distance into the ground, increasing the ground shock and crater. According to John Michener's calculations (Chapter 9), a penetration depth of only 5 m about doubles the crater radius. Such a doubling in radius could also be produced with a surface burst carrying 8 times the original yield, because the crater volume is proportional to the yield. So a half-megaton warhead that penetrates to 5 m is equivalent to a surface burst of 4 MT against superhard silos, and we can find the silo survivability against half-megaton earth penetrators (at 5 m depth) by looking at the 4 MT curve of Figure 3. We see that such half-megaton warheads could destroy superhard silos with high probability provided they also had MX-type accuracy. Guided re-entry would not be required.

The US has developed but not deployed a Pershing II warhead that is both terminally guided and able to penetrate to 30 m (21,23). However, the large weight penalty of this complicated warhead makes it impractical for long-range missiles. The Department of Defense is embarking on a high-priority effort to develop a new class of highly accurate earth penetrators to destroy hardened Soviet command and control posts buried deep underground (24).

The weight penalty might be reduced if earth penetrators dispensed with terminal guidance and penetrated only a few meters rather than the 20 or 30 meters desired against deep underground command posts. We have seen that just a few meters penetration, and MX-type inertial guidance, may be sufficient for a half-megaton warhead to pose a large threat to superhardened silos.

All in all, MaRVs and earth penetrators, separately or in combination, could be a reality for the mid-1990s, and would make superhard silos very vulnerable.

Superhard silo deployment makes little sense if the Soviets might deploy such warheads. A decision for superhard silos should be accompanied by arms control that prevents such developments. An intercontinental guided MaRV does not exist today, and would certainly need to be flight tested before it could be deployed (19). This development could be prevented by a flight-test ban on

MaRVs, provided the atmospheric maneuvering proves to be detectable. Earth penetrators could be prevented by a ban on the testing of nuclear weapons having a yield above one kiloton. Such a ban would prevent the development of a hardened primary charge for the weapon. Earth penetrators would require underground testing to assure that they will explode as planned following an encounter with hard rock at reentry speeds.

Superhard densepack?

It has been suggested that Midgetman or MX be deployed in a "densepack" of closely spaced superhard silos (25), similar to the closely spaced basing (CSB) once popular for the MX (23,26,27).

The idea behind CSB is to locate neighboring silos so near each other that attacking warheads destroy each other with fratricide effects (radiation, blast, fireball, turbulence, dust, debris) before they can destroy the silos. This idea seems particularly applicable to superhard silos, because the silos could be separated by as little as a few hundred meters without risking the simultaneous destruction of two or more silos by only one attacking warhead, and because very high-yield warheads are needed for the attack.

CSB has been controversial ever since it was proposed for MX basing, in 1982. One report states that "technically expert proponents and opponents of CSB hold such divergent views on the efficacy of fratricide for protecting CSB, (that) independent technical study of this issue may be necessary " (23). It is difficult to resolve this question because the atmospheric test ban prevents a direct check on the very complex theory.

Even without evaluating CSB's technical validity, two points seem clear: First, CSB is actually *more* vulnerable than ordinary spacing (several kilometers, for Minuteman silos) in at least one respect: Their very closeness makes the silos more susceptible to a "pindown" in which high-altitude bursts over the silo field could generate radiation, blast, or turbulence that would destroy missiles launched shortly after those bursts, preventing the launch of the missiles. If carried out by short-warning SLBMs, a pindown could prevent US ICBM launches during the time needed for ICBMs to arrive and to gradually destroy the entire closely spaced array.

Second, the main effect of CSB on Soviet thinking will be to enhance their uncertainty. As we argued in Chapter 3, uncertainty deters surprise attack but may do little to deter a damage-limiting pre-emptive strike in a crisis. In a crisis in which nuclear war seemed imminent, theoretical arguments about CSB and fratricide might do little to deter the Soviets from pre-empting against any forces that seemed vulnerable. SLBM pindown, and the disruptive effects of the first very large groundbursts, especially the effects on the array's command and control, make it at least plausible that CSB could be defeated.

Furthermore, CSB would give the Soviets further incentive to develop MaRVs and earth penetrators, for either development could defeat CSB. With higher accuracy, warhead yields could be smaller so that fratricide would be less significant. And underground explosions would eliminate many fratricide effects.

In fact, John Michener argues in Chapter 9 that cooperative effects from subsurface bursts could actually enhance the destruction.

Rather than contributing to stability, the uncertainties plus the threats from SLBM pindown, MaRVs, and earth penetrators, might make CSB less stable than normal spacing for superhard silos.

SLBMs and synergistic survivability

The Soviets are more likely to develop an ICBM capability against superhard silos than an SLBM capability. For one thing, it is more difficult to develop the needed MX-type accuracy for SLBMs than for ICBMs. For another thing, even with high accuracy, yields on the order of 4 MT are needed (assuming no MaRVs or earth penetrators), and such large warheads could probably not be MIRVed on SLBMs. Thus at least 500 missiles, over 30 submarines (at 16 missiles each), would be needed, just to attack the 500 superhard silos. Short of an all-out arms race, it is hard to imagine the Soviets doing thins.

Thus superhard silos would have the important effect of preserving the synergistic survivability of US ICBMs and bombers. As we have seen (Chapter 3), highly accurate Soviet SLBMs could make both ICBMs and bombers vulnerable to a single short-warning SLBM attack. Furthermore, mobile Midgetman may do little to solve this problem because mobile missiles are more vulnerable to SLBMs than to ICBMs (Chapter 7). Superhard silos work in the other direction, being essentially invulnerable to SLBMs.

Superhard silo survivability as a function of arms control

Although superhard silos will be vulnerable to a sufficiently determined Soviet military buildup and attack, the same can be said of mobile Midgetman, mobile MX, and probably every other feasible land basing scheme. Increased accuracy and large arsenals have precluded invulnerable land basing schemes. The best one can hope for is a *sizeable* (but not 100%) surviving fraction following any *plausible* (but not every conceivable) Soviet attack. But the plausibility of various Soviet attacks depends in turn on the numbers and kinds of missiles in the Soviet arsenal.

Thus we study the survivability of a US ICBM force comprising the present silo-based Minutemen and MXs plus 500 superhard-silo-based Midgetmen, under the four arms control scenarios described in Chapter 7: an *unconstrained arms race*, *SALT II limits*, *50% cuts* below SALT II limits, and *finite deterrence* (2000 strategic warheads on each side). We won't study superhard MXs, since we have already seen that they are vulnerable to small and plausible Soviet attacks. We assume throughout that the Soviets do not have guided MaRVs or earth penetratrators; we have already seen that superhard silos make no sense unless these developments are precluded by arms control.

In an *unconstrained arms race* the Soviets could clearly destroy the entire postulated US ICBM force. For example, with a doubled force including 600 accurate (100 m CEP) heavy ICBMs, they could destroy the 500 superhard silos with 500 single-warhead heavy missiles or with 170 3-MIRVed heavy missiles carrying 4 MT warheads. They would have plenty of warheads left to destroy the 1000 conventionally hardened MM/MX silos. They could also attack the strategic bomber bases, but most of the bombers would already be off the ground because of the 30-minute ICBM flight time. Alternatively, the Soviets might choose to attack US bombers with short-warning SLBMs, and they might also be able to destroy the MM/MX force in this way. But as we have noted, it is unlikely that SLBMs will be able to destroy superhard silos. Thus the Midgetmen could be launched following the SLBM attack and before the first Soviet ICBMs arrived. Although US ICBMs would not be independently survivable, the Midgetmen would be synergistically survivable with the bombers.

Under *SALT II limits* the Soviets would be limited to 308 heavy ICBMs, not enough for a single-warhead attack on 500 superhard silos. The most efficient attack would use 170 3-MIRVed heavy ICBMs, or 500 single-warhead medium ICBMs, carrying 4 MT warheads with 100 m accuracy. There would be plenty of Soviet warheads to also attack the MM/MX silos and the bomber bases, but as noted above the bombers should have time to escape.

With *50% cuts* the Soviets would not have enough heavy ICBMs for a 3-MIRVed attack on the superhard silos. For example, the Soviets could use 500 high-accuracy single-warhead medium ICBMs against the superhard silos, leaving a residual Soviet force of 200 or 300 ICBMs plus their SLBMs. This is certainly more stable than today's situation, because the attack would consume 70% of the Soviet ICBM force (and is thus less likely to be launched in the first place), whereas today the US ICBM force can be attacked by 15% of the Soviet ICBM force (or 30% of Soviet ICBM warheads). Again, synergistic survivability would be preserved.

Finally, under a *finite deterrence* scenario with 500 ICBM warheads (superhard Midgetmen on the US side), 500 SLBM warheads, and 1000 bomber warheads on each side, the entire Soviet ICBM force would be needed to attack the US ICBM force. The Soviets would come out "behind" in the warhead exchange, because some of their warheads would be duds and others would miss their targets. Although a damage-reducing pre-emptive strike is still conceivable, this arms control scenario is far more stable than today's situation, and more stable than the three preceding scenarios.

Summary and conclusions

US ICBMs could be housed in superhard silos of 7000-10,000 atm (100,000-150,000 psi) strength in the mid-1990s. Such silos could be destroyed only by including them within the crater of a nuclear explosion. For inertially guided warheads, MX accuracies (100 m) and large yields (4-20 MT) would be needed to destroy superhard silos with high probability. Such accuracies and yields are feasible on low-MIRVed heavy ICBMs or single-warhead medium ICBMs.

If the Soviets develop guided maneuvering warheads, or earth-penetrating warheads, much smaller yields (e.g. 0.5 MT) could suffice to destroy superhard silos. Thus superhard silos make no sense unless these developments can be precluded by arms control.

On inertially guided missiles, the required combination of high accuracy and high yield are probably not attainable on SLBMs. Thus superhard silos should be vulnerable only to ICBM attack. This has the important effect of preserving synergistic survivability of US ICBMs and bombers, and may be an important advantage of superhard silos over either today's silos-based MM/MX force, or a future mobile missile force.

Closely spaced basing for superhard silos is risky because of its uncertainties, its vulnerability to SLBM pindown, and its vulnerability to guided MaRVs and earth-penetrating warheads.

MX deployment in superhard silos is nearly as lucrative a target as highly MIRVed missiles in vulnerable silos today. Five hundred warheads on 50 superhard MXs could be destroyed by 50 single-warhead heavy or medium ICBMs if the attacking warheads have the necessary lethality.

Five hundred single-warhead superhard silo based Midgetmen, on the other hand, is a more stabilizing force *if* coupled with partial disarmament reducing Soviet arsenals by 50% or more. Under 50% cuts, a Soviet attack on the full US ICBM force would consume about 70% of their ICBM force. Even in an extreme crisis, the Soviets would probably not devote so large a fraction of their highly-valued ICBM force to this task. Under 80-90% cuts (finite deterrence), the situation would be even more stable.

Without partial disarmament, however, a US ICBM force including 500 superhard Midgetmen might not be much more stable than today's MM/MX force, because only a minor fraction of the Soviet ICBM force might be needed to attack it. Synergistic survivability would be maintained however, so that superhard silos would still offer this distinct advantage over a future ICBM force consisting either of conventionally hardened silos or of mobile missiles.

References and notes

1. William Arkin et al, "Nuclear Weapons," *SIPRI Yearbook 1985*, Taylor & Francis, London (1985).

2. Bruce A. Smith, "USAF reshapes SICBM effort," Aviation Week and Space Technology, 14 May 1984, p.20.

3. "Midgetman tests on basing alternatives," Jane's Defense Weekly, 24 Mar 1984, p. 427.

4. "USAF cites improved silos hardness capability," Aviation Week and Space Technology, 28 Oct 1985, p.85.

5. Edgar Ulsamer, "The prospect for superhard silos," Air Force Magazine, Jan 1984, pp. 74-77.

6. Edgar Ulsamer "Countering the Soviet strategic shield," Air Force Magazine, Oct 1984, pp. 72-77.

7. US General Accounting Office, *Report to Congress on UCBM Modernization*, GAO, Sept 1986.

8. Edgar Ulsamer, "Air Force systems command," Air Force Magazine, Aug 1984, pp. 53-61.

9. Kosta Tsipis, *Nuclear Explosion Effects on Missile Silos*, Center for International Studies, MIT, Cambridge, MA (1978).

10. Jonathan Medalia, *Small Single-Warhead ICBMs*, Congressional Research Service, Report No. 83-106F, Library of Congress, Washington, DC (1983).

11. Harold Brown, quoted in Steve Smith, "MX and vulnerability", ADIU Report, May/Jun 1982, p.3.

12. The result of the calculation is Eq. (4). If the reader wants to just accept this result, this section can be skipped.

13. Samuel Glasstone and Philip J. Dolan, *The Effects of Nuclear Weapons*, Departments of Defense an Energy, Washington, DC (1977).

14. George Ullrich, Shock and Strategic Structures Division, Defense Nuclear Nuclear Agency, quoted in Ref. 10.

15. Major James Jones, Defense Nuclear Agency (private communication).

16. Air Force Assistant Secretary Cooper, quoted in "SS-18 not capable of 250-foot CEP," Defense Daily, 22 May 1985, p. 121.

17. Barton Wright, *Soviet Missiles: Data from 100 Unclassified Sources*, Lexington Books, (1986).

18. Matthew Bun, *Technology of Ballistic Missile Reentry Vehicles*, Program in Science and Technology for International Security, Report No. 11, MIT, Cambridge, MA (1984).

19. Donald MacKenzie, "Missile accuracy--an arms control opportunity," Bulletin of the Atomic Scientists, Jun/ Jul 1986, pp. 11-16.

20. Aviation Week and Space Technology, 21 Sept 1987, p.13; also 7 Mar 1988, p.15.

21. Thomas Cochran, William Arkin, Milton Hoenig, *Nuclear Weapons Databook Volume 1: US Nuclear Forces and Capabilities*, Ballinger, Cambridge, MA (1984).

22. Robert Berman and John Baker, *Soviet Strategic Forces*, The Brookings Institution, Washington DC (1982), p. 57-58.

23. Jonathan Medalia, *The Effectiveness of the Proposed CSB Mode for Protecting the MX Missile*, Congressional Research Service, 25 Apr 1982.

24. "Earth-penetrating nuclear weapons," Aviation Week and Space Technology, 8 Jun 1987, pp. 28-29.

25. Bruce A. Smith, "USAF contractors defining SICBM," 6 Feb 1984, pp. 54-55.

26. Department of Defense, *White Paper on MX Closely Spaced Basing*, DOD, 12 Jul 1982.

27. Harry Wren, *Congress Rejects MX Dense Pack Deployment*, Congressional Research Service, Report No. 83-78 F, 19 Apr 1983.

Part V Chapter 12

Exploratory concepts for land-based missiles

David Hafemeister

Introduction

This book is primarily a search for a more stable and survivable basing mode for ICBMs in order to lessen the risk of nuclear war. A great deal of detail is given on options such as the Midgetman and MX missiles, based on mobile launchers or in super-hardened silos. This Chapter briefly reviews other concepts for modernization of ICBMs that would enhance their ability to penetrate possible future defenses and to attack extremely hard targets, and describes some of the conjectured defenses that might be deployed in the distant future to nullify the threat of ICBMs.

Defense suppression measures

In order to improve the ability of ICBMs to penetrate a defense, the attacking ICBMs can carry penetration aids that can nullify the sensors of the defensive systems (1), and can carry out a variety of other measures to circumvent the attacked nation's defenses. These include the following:

Radar masking. By releasing small strips of metal, called chaff, radar can be confused to think the target is elsewhere or much larger in size. The chaff can be released continuously, or ejected in various directions to look like individual reentry vehicles (RVs). Decoys, such as inflatable balloons made of aluminized mylar, will reflect radar, and look like an RV. The decoys are only viable during the mid-course phases. Figure 1 shows the MX Starburst concept, designed as a follow-on to the MX missile, that would allow for penetration aids and maneuvering reentry vehicle (MaRV) technologies.

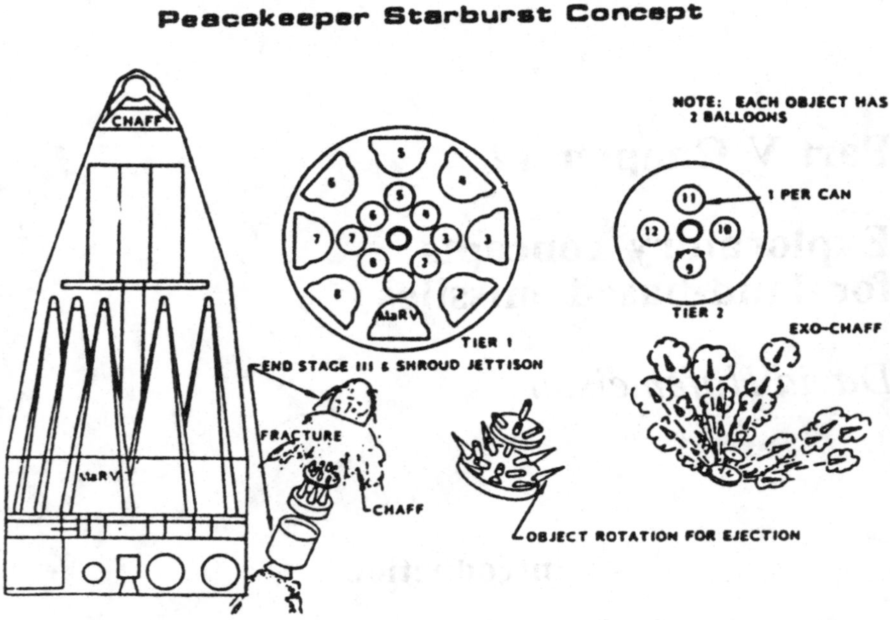

Figure 1. MX Starburst concept. Penetration aids can be added to the MX missile to overwhelm conjectured, futuristic defenses. Metallic strips, or chaff, shown just under the missile nose cone in the vertical cross section on left, give false reflected radar signals during the midcourse phase. The chaff is explosively ejected from just below the top of the missile shroud when the post-boost vehicle is above the atmosphere. Starburst has 56 inflatable balloons as midcourse decoys to spoof the defensive radars. The design allows for improving the accuracy of the reentry vehicles by making them maneuverable. The reentry vehicles are arranged in the outer circle of the post-boost vehicle. The design accommodates the weight of the chaff, decoys, and MaRVs by reducing the number of reentry vehicles from ten to eight. (Courtesy of the US Air Force.)

Infrared masking. There are a number of options for the offense to mask the infrared heat signal of an incoming RV from the defense. For example, aerosols can be released during the midcourse phase to hide the infrared. Or, one can confuse the defense by using decoy balloons or replicas with a modest heat source to make the decoy look like an RV. If the detectors cannot distinguish the RV signal from the decoy's signal, some of the defensive capabilities will be wasted by attacking the decoys. Positive detection of the RVs in the midcourse phase is very difficult whether one uses passive discrimination methods with infrared detectors or active discrimination with neutral particle beams. Alternatively, one can enclose the RV in a balloon, blocking the infrared signal and making the RV look like the decoy.

Attacking the defense. The attacking nation can always directly attack the defensive sensors (radars, infrared platforms, etc.) by directly attacking or blinding these systems with nuclear or other weapons. The attacking nation might pay a price in terms of the timing of events because the attack must destroy the sensors in advance of the attack. Of course, the defense would attempt to defend their sensors. Most likely, the attacking nation would use SLBMs which could arrive in 7-15 minutes, as compared to the 25-30 minutes for the ICBM. Alternatively, the offense can explode nuclear weapons high in the stratosphere to create a "beta blackout," i.e. free electrons that reflect the radar and hide the RVs.

DANASAT. If either the US or the USSR deployed a system of space-based interceptors, the other side might wish to destroy these defensive systems before launching its nuclear missiles. The attacking nation could create a hole in the pattern of space-based interceptors or sensors by using direct-ascent nuclear anti-satellites (DANASATs), or direct-ascent non-nuclear anti-satellite (DANNASAT) weapons to attack. Because 90% of the defensive systems are not over the ICBM fields at any given time, only about 10% of the systems would have to be attacked.

Depressed trajectories. Lowered trajectories allow an RV to fly beneath the horizon of early-warning radars and of exoatmospheric optical systems such as the space surveillance and tracking system of the SDI program. In particular, submarines can exploit this approach because they can compensate for the extra energy required to launch on a depressed trajectory by moving closer to their targets. The SS-N-6 submarine-launched ballistic missile is often described as a system which could come in under the ballistic missile defense sensors. A missile on a depressed trajectory remains in the atmosphere longer, and is more difficult to attack with the space based interceptors or with directed energy weapons. RVs flying depressed trajectories in the atmosphere will have greater thermal stresses because of the atmospheric drag. In addition their accuracy will be degraded, but a viable MaRV program could reestablish good accuracy.

Boost glide vehicle. Boost glide vehicles are launched from an ICBM, but then they glide through the atmosphere to the target, using global positioning satellites or stellar terminal guidance to fix their location. Some wing structures are necessary to supply lift. Their low radar cross section and low altitude make them more difficult to attack.

Fast burn ICBMs. Fast burn boosters that can climb to 80 km in 60 seconds would negate the attack on the boost phase by directed energy weapons (2). The extra fuel needed for the fast burn would diminish the payload, reducing the number of RVs for an MX from 10 to about 8. Boosters that burn somewhat more slowly are sufficient against a more modest defense of slower, space-based interceptors. Solid fuel ICBMs rising to 100-160 km in 100-150 seconds would greatly diminish the viability of the space-based interceptors. The maximum kill area of the guided interceptor rockets, nicknamed "smart rocks," is proportional to the square of the time t available to attack the boosting ICBM, $A_{max} = \pi v^2 t^2$. Because of the t^2 factor, boosting more quickly with solid-fueled SS-25s would greatly diminish the kill area for an interceptor. It can be easily shown that SBIs with flyout velocities of 6 km/s can cover only about 1% of the earth's surface. By using faster boosting systems, such as the SS-24 and SS-25, Cunningham (3) has shown that 20,000-100,000 space-based interceptors would be needed for an effective defense.

Smart post-boost vehicle. An ICBM's post-boost vehicle is the bus that carries and releases the RVs, while slightly maneuvering to improve the accuracy of the RVs. A post-boost vehicle could have sensing capabilities to be able to attack satellite systems and other ballistic missile defense platforms that might attack the post-boost vehicle.

ICBM pin-down. By using silos placed close together and hardened to 2000-7000 atm (30,000-100,000 psi), the defense could force the offense to lay down large warheads in a tight pattern. The closely-spaced blasts would then destroy many of the incoming RVs before they exploded, an effect known as "fratricide," so that many of the silos would survive. This idea was proposed for MX missile basing in 1982. In response to closely-spaced basing, the attacker could attempt to pin down the ICBMs with high altitude nuclear blasts and then gradually destroy the ICBMs in a slowly unfolding attack that avoids fratricide. For further details, see Chapter 11.

Recallable fast-escape systems. The recallable cruise missile and the fast escape recallable missile could be carried on airplanes. These systems would be recallable since the pilots can control the systems until launch, and the system could be further controlled after launch. The air-launched and sea-launched cruise missiles presently being deployed are not recallable once they are launched.

Enhanced attack technologies

New technologies to increase the ability of ICBMs to destroy hardened and other targets include the following:

MaRV. The present limit on ICBM accuracy is about 100 meters (Chapter 3). This amounts to a fractional accuracy of about 10^{-5} (CEP/range = 0.1 km/10,000 km). The uncertainty of the reentry phase is one of the largest items in the error budget. By using small glide wings, or by shifting the center of mass, it is possible to guide the RV closer to the target location. MaRVs have been under development since the beginning of the ICBM era. They have not yet been deployed on ICBMs, but were deployed on the intermediate range Pershing II ballistic missile with an accuracy of about 40 meters (4).

Midcourse updates. By shining a laser from an RV onto a reflecting satellite in a known orbit, or by using transmissions from global positioning satellites, it would be possible to update the RVs location and velocity, thus guiding an RV closer to its target.

Responsive targeting. By communicating target updates to an ICBM just before launch, it is possible to reconstitute an attack to make up for failed missiles, or missiles downed by the defense. This more efficient use of ICBMs is complex because of the very short times available and the need to protect the communications link to the ICBMs from being undercut by subversion.

Earth-penetrating warheads. Nuclear warheads could destroy silos by creating very large blast overpressures in the air. However, a warhead that explodes below the surface of the earth couples very strongly to the earth, making mechanical waves that can destroy by severe shaking. Some have conjectured that the Soviets have hardened their launch control centers and other command facilities, so the US is responding with the earth-penetrating warhead. These create an effective pressure about 20-50 times that of the conventional nuclear warheads exploded in the air. Earth-penetrating warheads could, when coupled with high-accuracy MaRVs or with cruise missiles, decapitate the leadership of the other side. In 1982, the earth-penetrating warheads for the Pershing IIs were cancelled (4). See Chapters 9 and 10 for further details.

Defensive technologies

A wide variety of defensive technologies have been considered. In 1987, the Defense Acquisition Board of the Department of Defense recommended the deployment of space-based kinetic-energy interceptors, but in June 1988 they considerably slowed the pace towards deployment. In the longer term, DOD is exploring the use of directed energy weapons. Some of the elements of these futuristic systems are as follows (Figure 2):

Surveillance and tracking systems. Satellites in orbit could provide data for an attack on the boost and midcourse phases. The boost phase of an ICBM would be monitored with infrared sensors located on the Boost Surveillance and Tracking System, while the midcourse phase would be monitored with the Space Surveillance and Tracking System with infrared and visible sensors, and perhaps with radar or laser radar sensors. The Airborne Optical System would deploy additional infrared and visible sensors. The reentry phase would be monitored with Terminal Imaging Radars. In the more distant future, neutral particle beams would be directed at RVs and decoys to create back-scattered radiation to actively discriminate between RVs and decoys.

SDI early deployment weapons. Space-based interceptors could be placed on carrier vehicles in orbit, with about 10 interceptors per carrier. The rocket propelled interceptors would have flyout velocities of about 6 km/second and infrared sensors would direct the interceptor to collide with the boost and post-boost phases of the missile. The midcourse would be attacked with kinetic energy weapons, the Exoatmospheric Reentry Interceptor System which is launched from the ground. The reentry phase would be attacked with the High Endoatmospheric Defense Interceptor that is also launched from the ground.

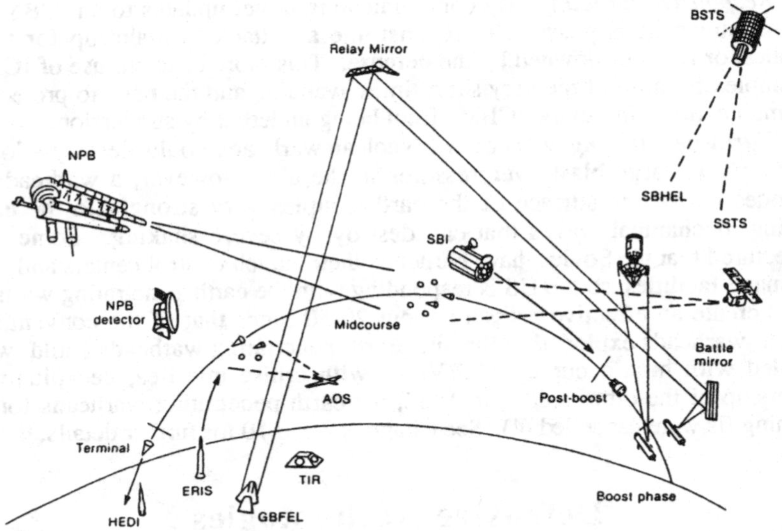

SDI sensor systems:

BSTS–Boost Surveillance and Tracking System (infrared sensors)
SSTS–Space Surveillance and Tracking System (infrared, visible, and possibly radar or laser radar sensors)
AOS–Airborne Optical System (infrared and laser sensors)
TIR–Terminal Imaging Radar (phased array radar)
NPB–Neutral Particle Beam (interactive discrimination to distinguish reentry vehicles (RV's) from decoys; includes separ
neutron detector satellite)

SDI weapons systems:

SBI–Space-Based Interceptors or Kinetic Kill Vehicles (rocket-propelled hit to kill projectiles)
SBHEL–Space-Based High Energy Laser (chemically pumped laser)
GBFEL–Ground-Based Free Electron Laser (with space-based relay mirrors)
NPB–Neutral Particle Beam weapon
ERIS–Exoatmospheric Reentry vehicle Interceptor System (ground-based rockets)
HEDI–High Endoatmospheric Defense Interceptor (ground-based rockets)

Figure 2. Major SDI sensors and weapons. (Courtesy of the Office of Technology Assessment.)

Directed energy weapons. A wide variety of directed energy weapons are being considered for deployment in the 21st century (2,5). Briefly, these systems include X-ray lasers on pop-up missiles based on submarines to attack the boost phase of the missile, free electron lasers and excimer lasers based on earth that would send their laser beams to mirrors at geosynchronous orbit and then onto battle mirrors in low-earth orbit to attack the boost phase, chemical lasers based in low-earth orbit to attack the boost or midcourse phases, and neutral particle beam weapons to attack the boost or midcourse phases.

References and notes

1. *Exploratory ICBM Concepts*, Ballistic Missile Office, Norton Air Force Base, CA, 1983.

2. *Report to the American Physical Society of the Study Group on Science and Technology of Directed Energy Weapons*, Reviews of Modern Physics, Volume 59, Number 3, Part II, Jul 1987.

3. C. Cunningham, *Energy and Technology Review*, Lawrence Livermore National Laboratory, Jul 1987, p. 16-17.

4. T. Cochran, W. Arkin and M. Hoenig, *U.S. Nuclear Forces and Capabilities*, Volume 1 of *Nuclear Weapons Databook*, Cambridge, MA, (1984).

5. *SDI: Technology, Survivability, and Software*, Office of Technology Assessment, Washington, DC, (1988).